Optical Communications
Rules of Thumb

John Lester Miller
Dr. Edward Friedman

McGraw-Hill

New York   Chicago   San Francisco
Lisbon   London   Madrid   Mexico City   Milan
New Delhi   San Juan   Seoul   Singapore
Sydney   Toronto

The McGraw·Hill Companies

Cataloging-in-Publication Data is on file with the Library of Congress

Copyright © 2003 by The McGraw-Hill Companies, Inc. All rights reserved. Printed in the United States of America. Except as permitted under the United States Copyright Act of 1976, no part of this publication may be reproduced or distributed in any form or by any means, or stored in a data base or retrieval system, without the prior written permission of the publisher.

1 2 3 4 5 6 7 8 9 0 DOC/DOC 0 9 8 7 6 5 4 3 2

ISBN 0-07-138778-1

The sponsoring editor for this book was Stephen S. Chapman and the production supervisor was Sherri Souffrance. It was set in ITC New Baskerville by J. K. Eckert & Company, Inc.

Printed and bound by RR Donnelley.

 This book is printed on recycled, acid-free paper containing a minimum of 50 percent recycled de-inked fiber.

Contents

Chapter 3   Electronics and MEMS

Chapter 4   Fiber Mechanical Considerations

Chapter 5   Free-Space Optical Communications

Chapter 8   Noise

Chapter 9   Optical Fibers

Chapter 10   Optical Signal Degradation

Chapter 11   Optics

Chapter 12   Splicing and Connectors

Chapter 13   Laser Transmitters

Chapter 14   Wavelength Selection

Preface

This book represents an extension of the authors' previous book, *Photonics Rules of Thumb,* published in 1996. This volume is intended to provide the telecommunications system designer with the same type of simple rules, references, and quick calculations that characterized the previous book. Herein we provide a quick history of how the authors started the series and its intent.

John Lester Miller's and Dr. Ed Friedman's careers intersected at the same place by very different paths. John spent some of his career in astronomy before joining the aerospace industry to work on infrared sensors for space surveillance and then search-and-rescue and telecom optical components. Ed spent much of his career working on remote sensing technologies applied to the Earth, its atmosphere, and the oceans and, more recently, astrophysical instruments and advanced pointing systems. John and Ed met in Denver in 1985, both working for a major government contractor on exotic electro-optical systems.

Both the telecom boom of the late 1990s and the military boom of the 1980s were halcyon days, with money flowing like water and contractors investing billions of dollars for some concepts that were overly optimistic or barely possible at best. In the center of the whole fray were semitechnical and nontechnical managers who were demanding systems capable of the impossible. The authors saw many requirements and goals being levied that were far from realistic and often the result of confusing interpretations of the capabilities of optical and electro-optical systems. At the same time, the potential for laser communications in all of its forms was evaluated by a large number of firms. The commercial success of the Internet led to a demand for bandwidth that could not be satisfied by copper systems, and designers invented the key technologies that enabled continuing increases in the bandwidth capability of installed fiber systems. Free-space communications got its share of attention, too, with applications ranging from satellite-to-satellite to building-to-building. In all cases, electro-optics technology formed the foundation of the advances that made the modern versions of these system concepts conceivable. In addition, many of the technologies that the authors were experienced with in the defense and science arenas (lasers, detectors, fibers, steering mirrors, etc.) found their way into communication systems. The reverse occurred, too. For example, space applications of fiber systems that allow distribution of signals on spacecraft with higher bandwidth, lower mass, and reduced cost made use of optical technology developed for terrestrial applications.

In both fields, communications and defense optics, the authors found a common ground when they discovered that many co-workers, in an attempt to outdo the competition, were promising to perform sensor and communications demonstrations that violated many rules of engineering, if not physics. In the late 1980s, Ed and John created a half-serious, half-humorous posting for the local bulletin board called "Dr. Photon's Rules of Thumb." Its content was a list of basic rules that apply when optics or electro-optics are being used.

Photonics Rules of Thumb contained 234 rules applicable to general electro-optics. For this update, the publisher and authors decided to break this book into two purer, separate editions. This book concentrates on optical telecommunications, and the other on electro-optical imaging and laser systems. We tried to keep the overlap to a minimum, but there are rules that are common to each volume (especially in the chapters on detectors and lasers). These rules simply point out fundamental concepts and are strongly valid and useful in both fields.

So, what is the value of a collection of rules? The answer lies in the way optical telecom and electro-optical designs are carried out these days. No one works alone. Commonly, a design study starts with a team meeting that includes representatives of all of the engineering disciplines that will work on the project. Next, a system engineer who has studied the customer requirements usually lays out a concept for the design team to improve and provide details. Wouldn't it be nice if each team member were prepared to detect fundamentally flawed approaches and identify and eliminate them early in the design effort? Then, team members each attack their part of the problem. Eventually, this leads to the inevitable confrontations when one subsystem designer (say the electrical designer) feels that some other part of the system (say the laser design) is too demanding and is forcing him to exceed the budgets of money, schedule, risk, and volume allocated to him. Most of these conflicted solutions could be avoided if each specialist knew just a little bit about the other systems that interact with his.

The intent of this book is to help the optics and E-O parts of those teams make quick assessments, using no more than a calculator (or at worst a spreadsheet), so that they don't have to try to support a fundamentally flawed concept. In addition, by providing simple rules that can be put to use by others on the team, even the non-optics designer can learn enough to protect the integrity of the process.

The book is also useful for managers, marketers, and other semitechnical folks who are new to the optical telecom industry (or are on its periphery) to develop a feel for the difference between the chimerical and the real. Students may find the same type of quick-calculation approach valuable, particularly in the case of oral exams where the professor is pressuring the student to do a complex problem quickly. Using these assembled rules, you can keep your wits about you and provide an immediate and nearly correct answer, which usually will save the day. After the day is saved, you should go back to the question and perform a rigorous analysis.

These rules are useful for quick sanity checks and basic relationships. Being familiar with the rules allows one to rapidly pinpoint trouble areas or ask probing questions in meetings. They aid in thinking on your feet and in developing a sense of what will work and what won't. But they are not, and never will be, the last word. It is fully recognized that errors may still be present, and for that the authors apologize in advance to readers and to those from whom the material was derived.

The quest for rules led to key papers, hallmark books, colleagues, and into the heart of darkness of past experience. Some of these rules are common folklore in the industry, whereas others have been developed by the authors. The book has been assembled from a wide variety of sources, including experts, existing textbooks, and technical papers. It is by no means comprehensive, but it does provide some key rules in a wide range of fields. Rule selection was based on the authors' perceptions of each rule's practical usefulness to telecom professionals in the early twenty-first century. The downselect was accomplished by examining every rule for truthfulness, practicality, completeness (e.g., are all the variables defined?) range of applicability, ease of understanding, and, frankly, ones that the authors just liked. As such, this is an eclectic assortment that will be more useful to some than to others, and more useful on some days than on others.

The authors also take note of the valuable function provided by professional organizations in the creation of knowledge and the dispersal of it to their members and other interested people. Professional journals and books proved invaluable in our search for these little nuggets of information. We have scoured the past (and recent) publications of SPIE, the Optical Society, the National Fiber Optic Engineers Conference, and related journals. In reality, though, we get the most recent and cutting-edge information from conference proceedings. In this regard, the National Fiber Optic Engineers Conference (NFOEC) is extremely valuable. Not only do they collect their proceedings on CD (don't we wish everyone would do that?) and provide the information in a form that includes sophisticated search tools, but their authors also graciously allowed us to use the assembled material. NFOEC was born in 1984 as a technical training meeting. It has grown every year since. Recent meetings have been attended by participants from 33 countries.

The authors certainly encourage our readers to attend (and contribute to) these types of meetings. They offer a wonderful opportunity to benefit from technical presentations, attend the trade show, network with colleagues, and even look for jobs. Each NFOEC Conference, for example, hosts thousands of professionals in the fiber optic industry, with hundreds of exhibiting companies plus numerous industry and financial analysts and publishers of technical journals and books.

This collection of rules and concepts represents an incomplete, idiosyncratic, and eclectic toolbox. The rules, like all tools, are neither good nor bad; they can be used to facilitate the transition of whimsical concepts to mature hardware or to immediately identify a technological path worth pursuing. Conversely, they can also be used to obfuscate the truth and, if improperly applied, to derive incorrect answers. The job was to refine complex ideas to simplified concepts and present these rules to you with appropriate cautions. However, it is your responsibility to use them correctly. Remember, "it is a poor workman who blames his tools."

Acknowledgments

There are many to whom the authors owe a debt; virtually every rule contained in this book is due to the specific contribution or the inspiration of someone in the field of electro-optics. Without their efforts to develop the technology and its many applications, this book would neither exist nor have any value. We also acknowledge the contribution made by the many reviewers who helped remove errors and improve the content:

- Mike Arden
- Dwight Dumpert
- Geoff Fanning
- Rolf Franz
- Andre Girard
- Philip Hobbs
- Ying Hu
- Bob Kaliski
- Dick Kerr
- Dennis McCall
- Chongjin Xie

We also thank all of those who suggested rules and provided permissions and all of those who helped review and improve our first book, *Photonics Rules of Thumb*. Finally, the authors recognize the role of our families (especially our beloved wives, Judith Friedman and Corinne Foster) for tolerating long periods of loneliness.

Introduction

Communication is the foundation of virtually all of mankind's enterprises, so it is not a surprise that, over the millennia, a great deal of time and energy have gone into developing its technologies. The twentieth century witnessed great advances in the speed and global coverage of communications as electronics and electromagnetic wave management both came to maturity, often to solve problems for the military but with significant collateral commercial impacts.

By the 1990s, the need for bandwidth to support all types of communication, but most importantly Internet, had become apparent, and the solution was at hand. Optical communications stepped forward and enabled the Internet (or was it the other way around?). As stated by Yu,

> Recent advances and innovations in optical technology have thrust optical networking onto center stage as the emerging underlying communications infrastructure supporting the Internet revolution. With its huge potential, the optical components and subsystems sector is well positioned to capture a significant portion of the economic tsunami rolling over the networking world.[1]

The difference between this new generation of communication and the developments that led to radio, television, and wire-based telegraph was the conjunction of electronics with optical technology to provide a capability not possible with RF or copper-based systems. In any case, there is no doubt now about the role that must be played by optical communication systems as the need for bandwidth continues almost without end. Any system that involves fixed point-to-point connections can benefit from its higher communication capacity, once a digital protocol has been adopted. Moreover, the advantages of the optical pathway have now been addressed by intersatellite and satellite-to-ground systems.

Performance is just one of the features that has made optical communication attractive. Cost is important, too, and here we find that optical systems (principally fiber) have no real competitor. Even in expensive-to-install systems like undersea, fiber has maintained its edge over more exotic solutions like satellite-based RF communications.

As participants in the application of optics to a number of industries, the authors felt compelled to extend their already successful concept—a book that captures simple rules that can be used by system designers in the optical communications realm. In so doing, we have selected about 250 rules that are thought to be easy to understand, easy to put into use, and able to guide designers through those topics in which they have limited expertise. These days, it is easy not to be an expert, since systems of all types (as well as communications) rely on the integration of widely disparate technologies, ranging from optics to electronics to thermal design, to ocean science, to geophysics, and so on.

One of the disturbing trends in high technology is the constant narrowing of the role of engineers. As the technology matures, it becomes harder for anyone working in an area of the electro-optics discipline to understand all of what is being done in related sciences and engineering. This book was assembled to provide engineers with a readily accessible overview of a wide range of optical telecom topics. There is no intent to compete with the stalwart texts or the many journals devoted to the optical telecommunication fields, all of which provide considerable detail in every area. On the other hand, this text is intended to allow any EO engineer, working in whatever specialty, to make first guesses over a wide range of topics that might be encountered in optical telecommunication system design or fabrication, as well to provide guidance in choosing which details to consider more diligently.

The technical evolution of optical telecommunications parallels, and feeds from, developments in a number of somewhat unrelated fields, including surveillance technologies, classical optics, nonoptical telecom, materials science, electronics, military research, and many others. The common thread of all of this effort, which really came into focus in the 1990s, is that scientists and engineers have been able to apply the highly successful modulated-optics technologies with the more ancient concepts and methods of millimeter wave and electronic communications. The merging of these fields has provided an unprecedented capability for communication at staggering data rates. Major departments at universities are now devoted to producing new graduates with specialties in this field. There is no end in sight for the advancement of these technologies.

According to Yu,

> Two major themes will progress in the optical components industry. The first will be a continuation of the current paradigm, where components will strive to push performance to maximize the transmission capacity and reach. This involves the push to develop higher bit rates and limiting OEO [optical-electrical-optical] conversion using ultra-long-reach optics and tunable components. The key here will be to understand the characteristics of both old and new types of fiber. The second theme is based on a new paradigm where the cost of fiber is no longer the dominant economic driver. Here, the key will be to understand the cost/performance trade-off that can drive new applications and markets. This new paradigm is just beginning to emerge. The creation of automation and design tools and the development of software tools for designing optical fiber/waveguide components and simulating system performance represent another emerging trend.[2]

The bulk of this book consists of rules, divided into chapters. Each chapter begins with a short background of the general subject matter to set the stage and provide a foundation. The rules follow. If there are additional references or other sources of the material presented in the rule, they can be found immediately following the rule. Because many rules apply to more than one chapter, a comprehensive index and detailed table of contents is included. The authors apologize for any confusion you may have in finding a given rule, but it was necessary to put them somewhere and, quite honestly, it was often an arbitrary choice between one or more chapters. Students and those new to the field will find the glossary useful. Here you will find definitions of jargon, common acronyms, abbreviations, and a lexicon intended to clarify confusing and ambiguous terms.

We have tried to be comprehensive and all inclusive in the selection of topics. For example, electronics lays the foundation for all of modern communications and gets attention in a number of chapters including Chap. 3, "Electronics and MEMS," and Chap. 8, "Noise." At the same time, optics are addressed by number of chapters including Chap. 5, "Free Space Optical Communications," Chap. 9, "Optical Fibers,", Chap. 10, "Optical Signal Degradation," and Chap. 11, "Optics." Fibers certainly play a critical role in the implementation of communication systems. Therefore, this topic gets plenty of attention in Chap. 1, "Cables and Conduits,", Chap. 4, "Fiber Mechanical Considerations," Chap. 7, "Network Elements," Chap. 12, "Splicing and Connectors," and Chap. 13, "Laser Transmitters." Finally, we have included two chapters that deal with system-level issues: Chap. 2, "Economic Considerations," and Chap. 6, "Network Considerations."

In selecting material for inclusion in the book, we have been somewhat "forward thinking" in that we have emphasized the future of optical communication (DWDM, higher bandwidth, and so forth). We have also included what, for today, is the state of the art in amplifiers, detectors, and others components. Among these are in-fiber components like gratings, amplifiers, and micromechanisms. We are well aware that, only a few years after the publication of the this volume, the technology will have greatly advanced, so there has been some urgency in our work to make the contents as up to date as they can be—for now.

At the same time, we have attempted to include down-to-Earth rules that help with things that aren't likely to change, since they derive directly from physics. These include divergence properties of Gaussian beams, noise of electronic components, and the physics-limited performance of detectors. While everyone is disappointed to discover that Mother Nature is the limiting factor in a design, it is comforting to know that those limits will not be moving anytime soon.

We have addressed the needs of the installer/technician by including topics on optical time delay reflectometry (OTDR) and the design and installation of conduits, management of thermal problems in junction boxes, and rules on the strength of cables as a function of their exposure to stressing environments.

A note to the reader: Both volumes in this new edition of ROT have the same set of underlying concepts—to provide quick resources that will likely benefit the experienced worker in the field by offering simple solutions to complex problems. When possible, we have included information that could be of use to the newcomer, but these examples are rare, since the foundations of knowledge required to have mastery of these complex topics takes years of training in the field itself, along with physics, mathematics, and other topics. Therefore, the reader should be prepared to do additional work in the books and professional journal libraries to gain additional background. Significant published journals include the IEEE's *Journal of Lightwave Technology* and *Photonics Technology Letters,* and OSA's *Applied Optics.* Also, the reader is referred to the slightly less technical trade such as *Fiber Optic Product News, Laser Focus, WDM,* and *Photonics Spectra.*

Alas, technology advances, and all of us wonder how we can possibly keep up. Hopefully, this book will not only provide some specific ideas related to telecommunications technology but will also suggest some ways of thinking about things that will lead to a whole new generation of such rules and ideas.

As we discovered with our book, not everyone has the same thing in mind when providing a rule of thumb. To qualify for our definition of a rule of thumb, a rule should be useful to a practitioner and possess at least most of the following attributes:

- It should be easy to implement.
- It should be able to provide a good approximation of the correct answer.
- The main points should be easy to remember.
- It should be simple to express.
- It should highlight the important variables while de-emphasizing the role of generally unimportant variables.
- It should useful insight to the workings of the subject matter.

In the previous edition of the book, we found it valuable to create a detailed standard form and stick to it as closely as possible, and we do so with this edition. However, we have simplified the format for this book to eliminate some duplication of material, which occurred with the more detailed structure of the previous edition. Thus, for this edition, the format has been simplified, grouping all additional information into the "Discussion" section of the rule.

The rules in this book are intended to be correct in principle and to give the right answer as an approximation. Some are more accurate than others. Readers who have to get a more precise answer can sometimes refer to the given reference of the rule, which usually contains some more detailed analyses that give an even better answer.

Of course, the information captured in this book is just a glimpse of what is available to provide the professional with information needed to remain current with new technological developments. The interested reader might want to explore the history of the field of fiber optics, a major component of the optical communications revolution, or other sources of the material presented in the rule. There are countless book titles on all of the elements of optical communications. Indeed, there are too many to allow any fair selection to be mentioned here, with more titles appearing weekly. A quick review of publishers' offerings should provide any professional in these fields to find reference and educational material. In addition, a large number of tutorials and reference materials exist on the Internet. Most are available for free and can provide insight into virtually any topic. For example, a number of web sites include information on optical networking that can be a great place to get started. Additionally, Lightreading.com, IMAPs.com, Lightbrigade.com, and the web sites of organizations such as the IEEE, OFC, OSA, SPIE, Telcordia, and EIA/TIA offer valuable information. Several telecom companies even offer class-like tutorials on the Internet. Not only does the Internet provide a survey of concepts and explanations of them, it has links that lead to manufacturers, professional organizations and free technical information. An additional resource is the growing list of universities that are publishing lecture notes for free. Leading this effort is MIT, which will publish all lecture notes from its classes by the end of 2002. Professional organizations also make information available, sometimes with a compulsory (free) registration, but the information is top-rank and broad in scope. For example, the International Engineering Consortium makes available a wide variety of information and, for the more serious student, provides a number of (for-fee) on-line courses. Finally, one should always consider manufacturers as a resource to be exploited. Their catalogs, whether on paper or the "net," will contain not only descriptions of their products but also information on the general underlying ideas and physics that make their products work. However, the reader of these should be diligent, as sometimes the truth may be bent to promote their particular product or network architectures that require their product. In summary, there is no end of information for the curious. Be wary that you don't spend all of your time seeking new information and forget to put your knowledge to work.

A warning to the reader is in order here. Unfortunately, a few of the references that inspired the rules use English units. Rather than make the conversion to metric units, we have left the units intact to indicate the preferred choice of the authors of those references.

References

1. J. Yu, 2001, "Emerging Themes for Next Generation Optical components and Subsystems," *Proceedings of the National Fiber Optic Engineers Conference*, Telcordia, p. 1623.
2. J. Yu, 2001, "Emerging Themes for Next Generation Optical components and Subsystems," *Proceedings of the National Fiber Optic Engineers Conference*, Telcordia, p. 1625.

1

Cables and Conduits

The topics of this chapter range from the sea bed to the home, yet one theme is retained; all of these rules relate to how cables are run, protected, and maintained. The first telecom fiber cable was not lit until April 22, 1977 (between Long Beach and Artesia), and it wasn't until 1988 that the TAT-8 crossed the Atlantic, yielding 40,000 good telephone connections (and over 1000 times provided by the first copper cable).[1] Today, for example, an advanced fiber transmitting 184 wavelength channels at 40 gigabits per second can carry more than 90 million phone conversations (enough to satisfy several teenagers).

One of the topics addressed herein is the management of cables in which large numbers of fibers are protected. Clearly, this is a critical topic in these days of constant demand for increased bandwidth, regardless of application. We also include the issue of allowing the field worker to recognize different fibers in these dense cables by use of color. Similarly, a number of rules relate to the problem of pulling cables through ducts and the size of the cables that can be accommodated. With increasing fiber density a common trend, we have included a number of rules that relate to this topic, including flat and tube-like installations.

The above topics lead directly to a set of rules related to the problems encountered when running large numbers of cables in underground conduits, particularly with respect to the potential for the collapse of the conduit. New conduit materials improve this situation, but care must be taken to ensure that the installation is consistent with local geophysics, keeping in mind that long runs will cause the conduit to encounter a variety of conditions. We also include a rule that deals with the thermal management of dense cables. Overhead cables get attention as well. We have included a rule that addresses the threat imposed by weather conditions at different locations in the U.S.A. as a result of wind and gravity sag. Another type of environmental threat comes from fiber usage in spaces where elevated temperatures are common and high enough to induce connector damage. One of the larger rules deals with the threat and properties of elevated temperature operation, the types of cables that are most susceptible, and other details. At the same time, humidity is a problem that cannot be ignored.

A number of the rules in this chapter relate to optical time domain reflectometry (OTDR) and its proper application in the diagnosis of cables and fibers. This is a particularly challenging topic when one is considering undersea applications. Another of the larger rules in the chapter deals with this topic. In addition, we have included rules related to the performance of OTDR systems, measured in terms of the accuracy of the location of fiber defects.

Two of the rules deal with the general properties of the signal-to-noise ratio that is desirable in dense cable systems. This includes some details about noise sources in household cable applications.

Finally, we have included several rules that deal with installation challenges. Aimed at avoiding reflections that threaten system performance, they include some details on common installation mistakes that should be avoided.

The reader will excuse the use of English units in some of the rules. They are popular with people working in some of the disciplines and were used in the original reference.

Reference

1. J. Hecht, *City of Light*, Oxford University Press, New York, p. 181, 1999.

INNER DUCTS

Groups of two to four 25-mm dia. ducts are commonly routed through 100-mm dia conduits.

Discussion

A key factor in the technical and financial success of underground fiber systems is the reliable installation and maximum exploitation of the duct work that goes into the ground. A major threat to these systems is excessive tension applied to the inner ducts. These are the tubes though which fibers are deployed and which are drawn through the conduit laid in the ground.

The standard for conduits is 100 mm diameter, although 150-mm units are becoming popular. In addition, directional drilling has allowed new flexibility in the installation of these conduits. Directional drilling complements the other installation methods, including trenching and installations above ground. In all cases, the driving factor in the evolution of the technology is the need to manage costs. Of course, a cost factor is the eventual performance of the installed system, since a failure of the duct or fiber inside can be very expensive.

As designs evolve, not only the installation factors and packing density are at issue, but also the selection of the materials used in the components. In addition to mechanical properties, duct work must exhibit suitable resistance to other environmental factors such as temperature, humidity and moisture, and (in above-ground applications) UV radiation. Persistent exposure to ozone can be a risk as well.

Reference 1 also points out that seals from section to section of the duct must be air tight to ensure that air-assist placement systems can be used and to ensure that debris and water do not enter the duct. Finally, it is obvious that uniformity of outside diameter and wall thickness is critical if the desired packing density is to be achieved.

A typical definition of packing density (P) is

$$P = \frac{w^2}{T_{ID}^2} \times 100(\%)$$

where w = ribbon stack diameter
   T_{ID} = tube inner diameter

References

1. R. Smith et al., "Selection and Specification of HDPE Duct for Optical Fiber Applications," *Proceedings of the National Fiber Optic Engineers Conference,* 1998.
2. J. Thornton et al., "Field Trial/Application of 432-Fiber Loose Tube Ribbon Cable," *Proceedings of the National Fiber Optic Engineers Conference,* 1997.

CABLE-TO-DUCT RELATIONSHIPS

1. The maximum diameter of a cable to be placed in 31.8-mm (1.25-in) inner duct is generally considered to be 25.4 mm (1.0 in).[1]
2. Historically, a maximum cable diameter of 25.4 mm (1.0 in) has been used as a "rule of thumb" for cable installations in 31.8-mm (1.25-in) ducts."[2]
3. 25.4- through 38.1-mm (1.0- through 1.5-in) inner ducts are commonly pulled as multiples of two to four ducts into 88.9-mm (3.5-in) square or 101.6-mm (4-in) round conduits."[3]

Discussion

Obviously, getting the maximum number of fibers into a duct is a good idea. Lail and Logan[2] also comment that a cable diameter of 25.4 mm (1 in) fills 64 percent of a 31.8-mm (1.25-in) duct.[2] It seems like, generally, a cable can be added to a duct if the cable does not exceed about 70 percent of the area of the duct.

Cables intended for installation in these inner ducts must have an outside diameter that is not only less than 1.25 in but also small enough to negotiate the bends and length of the conduit route. Until now, cable manufacturers have only met this specification of 1 in (25.4 mm) maximum diameter with cables containing 432 or fewer fibers. This, in turn, has limited service providers to 1296 fibers in a single 4-in conduit structure.

The small diameter of fiber optic cable compared to copper cables makes possible a means of multiplying the duct space. By placing multiple 31.8-mm (1.25-in) inner ducts in the existing 88.9- or 101.6-mm (3.5- or 4-in) conduit structures, the effective duct capacity can be increased by two or three times. The outside-plant challenge then becomes the inside diameter of the inner ducts rather than the number of available conduit structures.

References

1. E. Hinds et al., "Beyond 432 Fibers: A New Standard for High Fiber Count Cables," *Proceedings of the National Fiber Optic Engineers Conference,* 1998.
2. J. Lail and E. Logan, "Maximizing Fiber Count in 31.8-mm (1.25-inch) Duct Applications—Defining the Limits," *Proceedings of the National Fiber Optic Engineers Conference,* 2000.
3. R. Smith, R. Washburn, and H. Hartman, "Selection and Specification of HDPE Duct for Optical Fiber Applications," *Proceedings of the National Fiber Optic Engineers Conference,* 1998.

COLORED RIBBON CABLES

Colored ribbons provide a number of advantages in optical applications. They enable high fiber count, bulk fusing, and quick identification through the use of color.

Discussion

Colorization must be done properly, since the introduction of a coloring agent to the matrix material can "affect the cure performance, modulus, and glass transition temperature of the material."[1]

Clearly, the main advantage of a colored product is to reduce the time required to identify particular ribbons in the field. Figure 1.1 shows the advantage of colorization.

This new capability is not achieved without some cost. The peel and separation performances were verified through standard tests, including use of the midspan access peel kit and visual inspection of fiber surfaces after separation. Additionally, fiber-to-matrix adhesion has been quantified through the development and use of a high-resolution test method. This test measures the critical fracture energy of the ribbon matrix material. Data from this test method and the equation used are shown in Fig. 1.1. Through an understanding of the different process and material variables that control adhesion, the ability to manufacture ribbon with a specified adhesion value is obtained.

Reference

1. K. Paschal, R. Greer, and R. Overton, "Meeting Design and Function Requirements for a Peelable, Colored Matrix Optical Fiber Ribbon Product," *Proceedings of the National Fiber Optic Engineers Conference,* 2000.

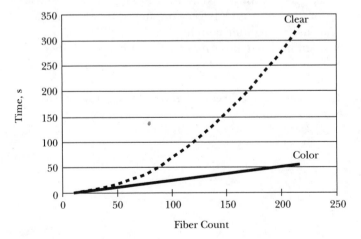

Figure 1.1   Recognition time as a function of fiber count.

TEST CABLES AFTER SHIPPING

Cables should be tested after shipping.

Discussion

Damage to cabling can occur during shipping or installation. Failing to test fiber cabling after it is delivered is a common mistake made by installers. This failure makes damaged cable detection difficult and returns awkward. An OTDR could be used in this case to shoot an optical profile on each fiber after the cable is received and still on the shipping reel. A permanent record will then be available for future use.

Failing to perform testing, verification, and documentation prior to the installation of the fiber end-termination equipment is a problem. If the fiber is not tested after installation, it cannot be determined whether it was installed correctly; serious equipment performance problems can occur. Furthermore, failing to document in the cable plant could make troubleshooting difficult later, as well as voiding warranty conditions of the installed network.

This reference[1] also suggests that test times can be reduced by varying the sampling rate as a function of fiber test length.

When testing short runs of fiber, there is typically not much information about the fiber available except for length and attenuation. Connectors and splices are generally not present, needed, or used for short lengths; new short runs would be reinstalled. The sampling rate of an OTDR will determine how much resolution the instrument has when capturing trace information. While it is important to maximize resolution for short distances, it is not mandatory for longer distances. Since it takes more time to take more sampling or data points, longer stretches of fiber can use a lower sampling rate, whereas medium lengths can use a medium sampling rate. This kind of incremental improvement in time helps when testing hundreds of fibers.

Reference

1. S. Goldstein, "Fiber Optic Testing Fiber to the Desk," *Proceedings of the National Fiber Optic Engineers Conference*, 1998.

END-OF-FIBER REFLECTIONS

End-of-fiber reflectance using OTDR can be computed as

$$R = Bns + 10 \times \log\left(H^{\frac{H}{5}} - 1\right) + 10 \times \log D$$

where R = reflectance of a pulse
Bns = fiber backscatter coefficient at 1 ns (a negative number)
H = height of the reflection with respect to the backscatter level
D = OTDR pulse width in ns (with some adjustment for the fiber attenuation over the pulse width and for pulse shape)

Discussion

When measuring the characteristics of a fiber cable with an OTDR, several measurements are typically acquired (e.g., splice loss, fiber loss per kilometer, distance to event loss, and so on). The referenced paper describes a method for improving OTDR measurements with uncertainties because the fiber backscatter coefficient is unknown.

The backscatter coefficient is generally a default value given by the OTDR manufacturer or entered by the operator, and it can be seen that this directly affects the error on the reflection value (e.g., a 1-dB error in Bns corresponds directly to a 1-dB error in the reflection).

Reference

1. F. Kapron, B. Adams, E. Thomas, and J. Peters, "Fiber-Optic Reflection Measurements Using OCWR and OTDR Techniques," *Journal of Lightwave Technology*, 7(8), 1989.

FIBER DENSITY

Flat ribbons can pack at least 40 fibers per 100 mm^2.

Discussion

The results shown in Fig. 1.2 assume a packing density of 70 percent, which is derived from experience with different interconnect designs. Figure 1.2 illustrates the density of fibers for different types of packaging.

The reference cautions that "in practice, the packing density would probably be even lower for the cables with larger bend radii, since they would be more difficult to manipulate in tight quarters."[1] The reference also provides the following table of information on different types of fibers.

	Units	12f flat ribbon	24f flat ribbon	2 × 12f ribbon, single-tube	250 µm, single tube	500 µm, tight buffer	Single-fiber cables
Height	mm	2.1	3.5	NA	NA	NA	NA
Width	mm	4.6	5.5	NA	NA	NA	NA
OD	mm	NA	NA	6.9	5.0	5.8	2.9
Area	mm^2	8.71	16.62	37.39	19.63	26.42	6.61
Fibers/mm^2	#/mm^2	1.4	1.4	0.6	1.2	0.9	0.2
Fibers/100 mm^2 (70% packing)	#/mm^2	96.4	101.1	44.9	85.6	63.6	10.6

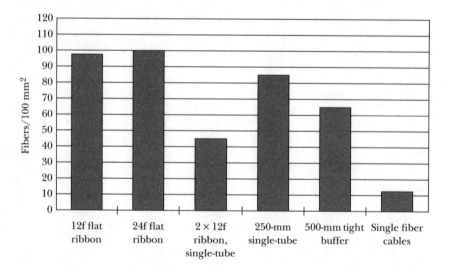

Figure 1.2 Fiber density for various packaging types.

Reference

1. J. Register and M. Easton, "Optical Interconnect Cabling for Next Generation Central Office Switching," *Proceedings of the National Fiber Optic Engineers Conference,* 2001.

GEOPHYSICS AND DUCTS

The force applied to the pipe by a gravel pocket can be crudely estimated from the bulk density of the pocket material (γ_g), the height of the pocket above the duct (H). The gravel pressure acting on the duct (assuming as dry, not saturated with water or slurry) can be estimated as follows:

$$\text{Earth Pressure } (P_E) = \gamma_g \times H/144 \text{ in}^2/\text{ft}^2 = 3.33 \text{ lb/in}^2$$

Discussion

New materials allow new services and capabilities. A good example is the use of 4- and 6-in high-density polyethylene (HDPE) duct, which has properties that allow it to be used in directional boring applications. These new technologies help manage cost and expand the range of applications that can be addressed. HDPE has the potential to assist in the directional boring process, where tensions can be tens of thousands of pounds. Moreover, forces on ducts can be substantial.

For example, consider a duct in a loose gravel condition. If we assume that the gravel falls and packs around the duct over a 6-ft length, the gravel soil resistance (F_S) may be estimated as

$$F_S = P_E \pi D_0 \, l \mu = 895 \text{ lb}$$

where D_0 = average diameter of the duct
  l = length of the pocket along the bore direction
  μ = coefficient of friction (est. 0.5) of the gravel at the duct surface

The referenced paper states it well. "It can easily be seen how combinations of buoyancy drag and earth resistance can escalate quickly to challenge the tensile yield strength of the duct."

Reference

1. R. Smith, R. Washburn, H. Hartman, "HDPE Duct Selection and Specification of HDPE Duct for Optical Fiber Applications," *Proceedings of the National Fiber Optic Engineers Conference*, 1998.

CABLE DEFLECTION

The actual deflection of a cable under compressive load can be approximated as follows:

$$\delta = \frac{3\pi KPD_J^3 D_T^3}{4\left[E_T t_T^3 D_J^3 + E_j t_T^3 TD_T^3\right]}$$

where δ = cable deflection
D_J = average jacket diameter
P = compressive load applied in force per unit length
E_T = tube flexural modulus
E_J = jacket flexural modulus
K = empirically determined constant
D_T = average tube diameter
t_T = tube wall thickness

Discussion

It is typical to design a tube to be strong enough to keep the deflection of the cable below a desired level when the cable is subjected to the rated compressive load. Given the inside diameter of the tube and the height of the ribbon stack, the desired level of deflection is easily determined. Basically, the tube should not deflect to the point at which it imparts stress on the ribbon stack. If the ribbon stack does deform from deflection, the fibers could be stressed to the point of attenuation, and the long-term reliability of the fibers could be jeopardized.

We also note that the maximum desired deflection is

$$\delta_{max} = D_i - \sqrt{(Nh)^2 + w^2}$$

where D_i = inside diameter of the tube
N = number of ribbons in the stack
h = height of an individual ribbon
w = width of an individual ribbon

Reference

1. M. Ellwanger, S. Navé, H. McDowell III, "High Fiber Count Indoor/Outdoor Family of Ribbon Cables," *Proceedings of the National Fiber Optic Engineers Conference*, 1996.

NONMETALLIC CABLE STRENGTH

The relationship the between tensile force P of cable installation with cable tensile strain ε and tensile member outer diameter D in the elastic region is given by

$$P = \varepsilon\pi\left[E_1\frac{D^2}{4} + E_2\left(\frac{R^2}{4} - \frac{D^2}{4}\right)\right]$$

where E_1, E_2 = Young's modulus of the central tensile member and the slotted rod
R = cable outer diameter

Discussion

There is understandable motivation to create fiber cables that carry large numbers of strands without the necessity to include steel for strength. It is widely known that the Young's modulus of Kevlar® fiber-reinforced plastic (FRP) is much lower than that of steel. This fact means that, when FRP is used as the strength member in a 1000-fiber cable, the FRP component must be larger than the steel counterpart, which has an adverse impact on the bending flexibility of the cable.

Reference

1. H. Iwata, M. Okada, S. Tomita, and N. Kashima, "Nonmetallic 1000-Fiber Cable Using PBO FRP," *Proceedings of the National Fiber Optic Engineers Conference*, 1997.

SWELLING RESULTING FROM CLEANING OF MATRIX MATERIALS

This rule is intended to alert users about swelling that can occur when the fibers and fiber bundles are cleaned using typical hydrocarbons such as toluene, acetone, ethyl alcohol, isopropyl alcohol, or light mineral oils. The cleaning material penetrates into the fiber coating or matrix material, and the entire bundle swells.

Discussion

Once the material reaches equilibrium, the swelling ratio, q, can be estimated from the following equation:

$$ q_m^{5/3} = \left(\frac{v_s M_C}{v_1} \right)\left(1 - \frac{2 M_C}{M} \right)^{-1}\left(\frac{1}{2} - \chi_1 \right) $$

The parameters are the ribbon matrix: specific volume, v_s, average molecular weight of a ribbon matrix cross-link, M_C, and overall polymer molecular weight, M. The parameters for a given interactive liquid are the polymer/liquid interaction parameter, χ_1 and the molecular volume, v_1. For fiber coatings and ribbon matrix material, all liquids with χ_1 values of nearly 0.5 will cause no appreciable swelling, but χ_1 can be expected to take on a range of values and result in dramatic swelling in some cases. The equation applies to low to moderate cross-link density material.

The quest to achieve higher packing density of fibers has led to a number of ideas for improved packaging. Ribbons of multiple fibers are often further packaged by using resins and a strength member.

Reference

1. D. Rader and O. Gebizlioglu, "Ribbon Matrix Materials Reliability: Compatibility with Cleaning Media and Cable Filling Compounds," *Proceedings of the National Fiber Optic Engineers Conference*, 1997.

OPTICAL TIME DELAY REFLECTOMETRY (OTDR)-BASED OPTICAL RETURN LOSS (ORL)

During OTDR, backscatter and reflection signals can be approximated as

$$\frac{\eta}{\beta v}(e^{\beta v T} - 1) \approx \eta T$$

where β = attenuation coefficient in km^{-1}
η = backscatter parameter (watts/joule)
T = pulse width in seconds
v = velocity of light in fiber (c/index of refraction) in km/s

Discussion

This simplification works when $\beta v T$ is $\ll 1$. In that case, the exponent on the left of the equation can be expanded to yield $[(\beta v T + 1) - 1]$. The approximation holds when the exponent is less than about 0.1 and over a wide range of conditions when short pulses are being considered. For example, for typical fiber index of refraction, v is about 205,000 km/s. Thus, for a pulse of 1 µs, the attenuation can be as high as 0.5 km^{-1} and still satisfy the conditions. This high attenuation coefficient is not likely to be found in a typical fiber.

The received backscatter power, P_b, as a function of distance z for a square input pulse of power P_0 is

$$P_b = P_0 \eta T e^{-2\beta z}$$

Reference

1. T. Murphy and N. Santana, "Accuracy of OTDR-Based ORL Measurements," *Proceedings of the National Fiber Optic Engineers Conference*, 1997.

OUTAGE RATES

One can expect the following causes and outage rates:

	Aerial cable (% of failures)	Buried cable (% of failures)	Underground cable (% of failures)
Dig-ups	0	84	94
Collision	45	0	2
Fires	10	0	0
Gunshot	18	0	0
Lightning	2	2	0
Rodents	6	10	2
Floods/wind/ice	19	4	2

Discussion

The above table is from Ref. 1. Obviously, the cable design and protection is an important factor in its resistance to natural caused outages. It is interesting to note the large number (over 25 percent) of outages caused by natural events such as lightning, rodents, floods, wind, ice, earthquakes, and hurricanes. Chamberlain and Vokey[1] give the following tables to use for a quick relative comparison of cable resistance to natural outages.

Cable type	Lighting damage	Power station	Rodent damage	Cable locate	Sheath maintenance	Water permeation	Cost performance
Metallic armored	G	F/P	E	E	E	E	E
All dielectric w. GRP armor	E	E	G	P	P	G	F/P
All dielectric w/o armor	E	E	P	P	P	G	G

E = excellent, G = good, F = fair, P = poor, GRP = graphite reinforced plastic.

Locate wire, AWG	Resistance, ohms/mi	Attenuation, dB/mi	Max. locate, mi
24	136	2.96	15
22	85.2	2.52	19
19	42.5	2.0	25
14	13.3	1.45	38
10	5.27	0.97	53

It is also interesting to note that the vast majority of outages can be stopped by simply planning properly (dig-ups) or making telecommunications networks more resistant to collisions and natural disasters. Not only do rodents eat a substantial amount of the calories grown for human consumption, they can cause serious outages and damage as they chew through the fiber cables. More importantly, in the case of a dig-up or a rat bite, one cannot locate the damage without an OTDR (optical time delay reflectometry) instrument. And even this instrument needs to be field positioned and exercised, which increases the outage time. As pointed out in Ref. 1,

> By far the greatest cause of major outages for buried cable is due to dig-ups. The FCC's Network Reliability Council report states that dig-ups account for nearly 60 percent of failures. Call-before-you-dig or one-call programs are strongly promoted to reduce the high rate of outages. In support of these programs, new toning and cable locating systems are available which can reach up to 50 miles or more. These systems include a rack mounted generator, which can be dialed up and controlled remotely, along with highly sensitive and selective hand held receivers which provide precise cable location and depth information, thereby minimizing the risk of mislocates. These cable tracing systems, which are by far the most effective means to locate and mark both buried and underground cable, require a metallic conductor to operate over. A metal armor is very well suited for this purpose.
>
> Where all dielectric cables are used below ground, an additional insulated conductor must be placed along with the cable to provide some means of tone locating. There are several drawbacks to this approach. The separate conductor adds cost. The trace conductor must be precisely located next to the dielectric cable or mislocates result, and any damage to the trace conductor can interrupt the locate signal. Due to the resistive and shunt reactance losses, the trace conductor must be 10 AWG or larger to be effective . . . .

Obviously, these rates do not include failures from hardware, as they are not included in "outages" but instead are classified as failures. That said, the quicker the repair the better. This harkens to modular network elements and fast cable repairs. More important is the ability to rapidly switch between rings or paths to keep all or most of the service alive when an outage occurs. The restoration time can be crudely estimated from Fig. 1.3, derived from Ref. 2.

References

1. J. Chamberlain and D. Vokey, "Metallic Armored vs. All Dielectric Fiber Optic Cable, the Pros and Cons," *Proceedings of the National Fiber Optic Engineers Conference*, 1998.
2. J. Nikolopoulos, "Network Planning Considerations Associated with Large SONET Ring Deployment," *Proceedings of the National Fiber Optic Engineers Conference*, 1998.

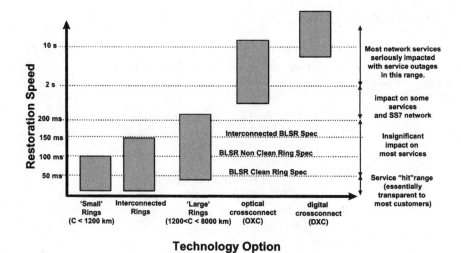

Figure 1.3   Impact of network architecture on restoration speed (from Ref. 2).

POSITION ACCURACY

Anyone who uses optical time-domain reflectometry (OTDR) wants to determine the accuracy of the indicated positions of defects found in the fiber. The position uncertainty can be expressed as

$$\Delta L = \frac{c}{2n}\left[W^2 + \left(\frac{0.44L}{F}\right)^2 + \left(\frac{0.44}{B}\right)^2\right]^{1/2}$$

where W = pulse width
  F = bandwidth per unit length of fiber
  ΔL = uncertainty in the continuous fiber length
  L = length of the fiber
  B = bandwidth of the measurement
  n = index of refraction

Discussion

From this rule we see that, as pulse length increases, the accuracy of the OTDR decreases (ΔL gets bigger). This rule seems to work for both conventional and graded index fibers. This is particularly important for shorter cables, as might be the case with plastic optical fibers (POFs).

Reference

1. T. Sugita, "Optical Time-Domain Reflectometry of Bent Plastic Optical Fibers," *Applied Optics*, 40(6), Feb. 20, 2001.

OTDR Measurements in Systems with Optical Amplifiers

The presence of optical amplifiers changes the way in which time delay reflectometry can and should be applied.

Discussion

The referenced paper shows that, for a bandwidth of $1/\Delta T$, one can compute the number of detected photons (N) as

$$N = \frac{P_{refl}\Delta T}{h\upsilon} = \frac{1}{2}\gamma_R S\Delta T^2 P \cdot e^{-2\alpha L}v_g$$

where $h =$ Planck's constant
  $\upsilon =$ optical frequency $= c/\lambda$
  $P_{refl} =$ reflected power
  $P =$ power projected into the fiber
  $\Delta T =$ duration of the pulses sent into the fiber
  $L =$ distance down the cable
  $\alpha =$ attenuation in the cable (a typical number is 0.00004343 m^{-1})
  $S =$ fraction of the scattered light captured in the backward direction [equal to $(NA)^2/n_I$ where NA is the numerical aperture of the fiber and $n_I =$ index of refraction of the fiber core]
  $\gamma_R =$ Rayleigh backscatter coefficient (a typical value is 0.7 km^{-1})
  $v_g =$ group velocity

Reference

1. L. Pedersen, "OTDRs in Systems with Optical Amplifiers," *Proceedings of the National Fiber Optic Engineers Conference*, 1996.

OTDR SNRs

It is common to describe the tolerable noise in an OTDR system as being twice the noise standard deviation.

Discussion

Using this definition, the distance from the event to the noise floor is relatively easy to measure. The variance of the local waveform noise is

$$\sigma_{db}^2 = \frac{1}{N}\left[\sum_i \{[5\log(w_i + n_i)] - 5\log(w_i)\}^2\right]$$

where w_i = linear waveform data

n_i = linear noise added to the linear waveform data by the OTDR's acquisition circuitry and optical receiver

N = number of samples taken

If we define the height of the pulse above the noise floor as δ,

$$\delta = 5\log(w/2\sigma)$$

(where σ is the standard deviation of the linear noise) and take the first two terms of the Taylor's expansion of the logarithms, we get

$$\sigma_{db} = 1.0857 \times 10^{-0.2\delta}$$

Reference 1 suggests that a figure of merit for distance measurement accuracy is the average event slope divided by the standard deviation of the local waveform noise (in the logarithmic domain),

$$F = m/\sigma_{dB}$$

where m = average event slope

Reference 2 shows that the signal-to-noise ratio is determined by the variance in the number of modes. It shows that

$$SNR = \frac{\text{Total number of photons}}{\sqrt{\text{Variance in the total number of photons}}}$$

$$= \frac{Mq}{\sqrt{Var(Mq)}} = \frac{Mq}{\sqrt{MVar(q)}}$$

This is a standard definition of SNR (see related rules in Chap. 8 of this book), but it includes the concept of the number of photons per mode, $q = N/M$ where M is the degree of coherence of the transmitted pulses, and N is the number of photons, as computed above.

For coherent light, $Var(q)$ becomes equal to q since photons follow a Poisson statistic, but in the case of a typical OTDR laser, the coherence is limited, and $Var(q)$ becomes equal to $q(q + l)$. This gives an expression for SNR of

$$SNR = = \frac{Mq}{\sqrt{Mq(q + 1)}} = \sqrt{\frac{ME}{E + h\upsilon M}}$$

where E = the energy detected in the bandwidth $1/\Delta T$

References

1. D. Anderson, "Multi-acquisition Algorithms for Fully Optimized Analysis of OTDR Waveforms," *Proceedings of the National Fiber Optic Engineers Conference,* 1996.
2. L. Pedersen, "OTDRs in Systems with Optical Amplifiers," *Proceedings of the National Fiber Optic Engineers Conference,* 1996.

THERMALLY INDUCED BUCKLING

If r is the fiber radius and l (the unsupported length of a fiber) is large enough, compressive stresses generated by cooling can cause buckling of the fiber according to the following equation:

$$\alpha T = \frac{5\,r^2}{l^2}$$

where α = the cable's structural backbone thermal expansion coefficient
         T = temperature drop

Discussion

Since the structural stiffness of the fiber is much less than that of the cable backbone, the latter essentially expands or contracts freely and controls the displacement of the fiber. Differential strain is generated by temperature changes because of the difference between thermal expansion coefficients of the fiber and cable backbone.

The buckling length for 125-µm dia. fiber with a temperature drop of approximately 100°C is approximately 3 mm, at which point tensile stress induced by buckling can cause fracture. With unsupported fiber of this length or greater, cooling from the epoxy cure temperature, or thermally shocking the connector between temperature extremes, can cause buckling and result in eventual failure unless there are mechanisms to relieve the compressive stress. The expansion coefficient for the example is about 20×10^{-6}. The calculation also assumes that the fiber is rendered immobile at both ends of the unsupported section.

References

1. B. LeFevre et al., "Failure Analysis of Connector-Terminated Optical Fibers: Two Case Studies," *Journal of Lightwave Technology*, 11, 537–541, 1993.
2. B. LeFevre et al., "Analysis of Fiber Fracture in Connectors," *Proceedings of the National Fiber Optic Engineers Conference*, 1998.

TEMPERATURE-INDUCED CABLE LOSS

Temperature-induced cable loss (TICL) in fiber optic cables can be severe—on the order of multiple decibels. Optical loss can be 10 dB or higher over a temperature range from 15 to –40°C.

Discussion

The references[1-4] offer the following details:

- Loss occurs at low temperature.
- There is about 10 times as much loss at 1550 nm as at 1310 nm.
- Loss occurs mostly in the range of 3 to 10 m from a termination.
- TICL is seen in cables that have been exposed for at least one summer, and loss increases with further aging.
- Proper termination of the central member is critical and can determine the amount of TICL that is experienced.
- Watch for large variations in fiber loss from one fiber to the next within the same given cable. The variations increase with the number of thermal cycles.
- TICL is more pronounced at 1625 nm than at 1550 nm, which provides a sensitive method of detection of TICL.[2,3] For example, effects not detectable at 1310 or 1550 nm can be detected by 1625-nm measurements.

It is noted that cables contain elements of widely varying thermal expansion coefficients and long-term dimensional stability. Glass is generally dimensionally stable and has a low coefficient of thermal expansion (CTE). Conversely, most polymers experience some irreversible dimensional change upon heating and thereafter have high CTE.

References

1. G. Kiss et al., "New Developments in Temperature-Induced Cable Loss: Scenarios, Materials, Single Fiber TICL, and 1625-nm Monitoring," *Proceedings of the National Fiber Optic Engineers Conference,* 1997.
2. S. Duquet and E. Gagnon, "Transmitting and Testing in the 1625-nm Window: Panacea or Pandora's Box," *Proceedings of the National Fiber Optic Engineers Conference,* 1998.
3. http://www.acterna.com/downloads/white_papers/Fiber-Optic/ mtse_1625nm_wp_ae_0202.pdf, 1625-nm requirements, February 2002, www.acterna.com.
4. O. Gebizlioglu and G. Kiss, "Temperature-Dependent Performance of Loose Tube Cables: Buffer Tube Materials and Cable Structure Issues," *Proceedings of the National Fiber Optic Engineers Conference,* 1996.

GROWTH IN UNDERSEA FIBER CAPACITY

Historically, the growth in undersea fiber capacity follows an exponential curve with a ten-fold increase about every three years.

Discussion

Undersea communication continues to compete well against satellite systems. Part of the its success stems from the continued evolution of achievable performance that offsets the high cost of laying the cable. Figure 1.4 illustrates trends in capability per fiber pair. Laboratory results illustrate that additional growth in capability can be expected as new technologies are introduced.

Reference 1 shows that the current implementation of undersea cables has the attributes described below. They expect the next generation of undersea fiber optic networks to include technology that will expand the traffic-carrying capacity on each fiber pair to 160 Gb/s. With up to six fiber pairs in the undersea cable, the cross-sectional cable capacities will reach 960 Gb/s.

1. Currently, repeaters support up to four fiber pairs using erbium-doped fiber amplifiers (EDFAs) that are pumped with 1480-nm lasers. The next generation of cables will support up to six fiber pairs using EDFAs being pumped with 980-nm lasers. The 980-nm pumps allow EDFAs to be reliably built with higher output power, a lower noise figure, and wider optical bandwidth than those using 1480-nm pumps. Future systems may also have Raman amplifiers.

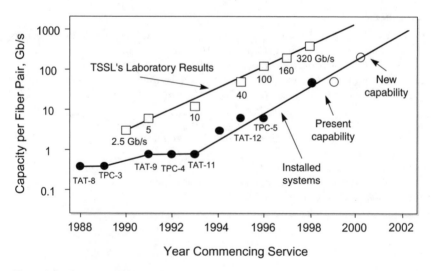

Figure 1.4   Capacity per fiber pair.

2. Currently, systems employ 10-nm usable optical bandwidths over transoceanic distances using gain equalization filters along the transmission path. This allows up to 16 channels on each fiber, with each channel operating at 2.5 Gb/s over distances as long as 12,000 km. The next generation will use optical bandwidths around 14 nm wide maintained over transoceanic distances by periodically deploying gain equalization flattening along the fiber path. These systems support 16 channels per fiber pair, with each channel operating at 10 Gb/s over distances of 10,000 km.

3. Current systems use both non-zero dispersion-shifted fiber (NZDSF) and nondispersion-shifted fiber (NDSF) in about a 10:1 ratio, respectively. The NZDSF's dispersion is about –2 picoseconds per nanometer-km (ps/nm-km) at 1.55 μ, and the NDSF's dispersion is about +17 ps/nm-km at 1.55 μ. The properties of these two fiber types result in a balance between performance and concerns about nonlinear effects. The authors expect future systems to also include large mode fiber (LMF). These fibers are also referred to as *large effective area fibers*. The LMF helps to minimize the additional impairments resulting from nonlinearities in 10 Gb/s carrier channel systems due to its increased effective area.

4. High-performance terminal equipment specifically designed for transmitting and receiving dense wavelength division multiplexing (DWDM) carrier channels for undersea transmission is already in use. This technology employs forward error correction (FEC), synchronous polarization scrambling, signal pre-emphasis, and dispersion compensation.

5. The current system employs branching units located off the continental shelf that support fiber routing by splitting fiber connectivity between the main undersea fiber optic trunk cable and a branch cable terminating at a landing site along the cable route, or branching units that use wavelength selective filtering to split the capacity between the main trunk cable and the branch cable.

6. Today's network systems are designed to be self-healing by exploiting equipment redundancy, facility protection, or a combination of both.

7. They also employ a transmission topology that is overlaid by active security measures to ensure fault, performance, and security management features.

Reference 2 provides an expression for SNR of submarine cable systems.

$$\text{SNR (dB)} = 58 + P_{\text{out, channel}} - \text{Loss}_{span} - N_F - 10 \log_{10} N$$

where   SNR = signal-to-noise ratio (in dB)

$P_{out,channel}$ = output power per channel from the amplifiers

$Loss_{span}$ = span loss, including fiber attenuation and splices
NF = noise figure of the EDFAs
N = number of EDFAs

The equation implies that a large number of low-gain amplifiers provide better SNR than a small number of high-gain amplifiers.

References

1. W. Marra and P. Trischitta, "Dense WDM Undersea Fiber Optic Cable Networks," *Proceedings of the National Fiber Optic Engineers Conference,* 1998.
2. T. Atwood, "Designing a Large Effective Area Fiber for Submarine Systems," *Proceedings of the National Fiber Optic Engineers Conference,* 1999.

WIRES IN THE HOME

Once a *very high speed digital subscriber line (VDSL)* signal enters the home, a myriad of interferences are still possible.

Discussion

The reference reports that there have been times when plain old telephone service (POTS) and Ethernet services were provided in the same cable sheath as VDSL and were found to be problematic. POTS produces very infrequent ring trip interference. Ethernet, on the other hand, is a Manchester encoded signal that is very broad spectrally and will definitely interfere with VDSL. Weeks also expects that "emerging HPNA signal may also interfere with VDSL as it is positioned on top of the VDSL spectrum."[1] The authors are referring to the standard developed by the Home Phoneline Networking Alliance, which uses existing telephone lines to create network resources within a home.

Reference

1. W. Weeks, "Real World Performance of FTTN/VDSL Systems," *Proceedings of the National Fiber Optic Engineers Conference,* 2000.

CABLE SAG

Final sag is about 2.5 times the initial sag.

Discussion

Installers and operators have to worry about the ability of cables to withstand loads encountered in the environment, such as snow, birds, and squirrels. Reference 1 notes that cables are under typical tensions of 2000 to 2500 lb, independent of the maximum operating tensions. All dielectric self-supporting (ADSS) cables will sag significantly under load.

It is also important to recognize the role that weather plays in defining the probable loads of overhead cables. Wind and ice are the principal culprits. A map of the severity of overhead cable loading by ice, wind and snow can be found in Ref. 2.

References

1. S. Woods, "ADSS Myths Dispelled: The Truth about Sag and Tension," *Proceedings of the National Fiber Optic Engineers Conference,* 1998.
2. *Mechanics of Aerial CATV Plant,* Technical Note/1006-A, Times Fiber Communications, Inc., © September 1995. The document can be found at http://www.timesfiber.com/techpdf/1006a-tn.pdf.

BRILLOUIN PHENOMENA

Brillouin phenomena scatters backward along a fiber, and Raman scatters forward; both tend to scatter to longer wavelengths.

Discussion

Both of these phenomena can cause problems, with Brillouin as a backflow to the switches and transmitters. Raman scatting is generally weak but can add to the noise floor.

MAXIMIZING FIBER COUNT IN A DUCT

A cable can generally be added to a duct if the cable does not exceed about 70 percent of the area of the duct.

Discussion

Obviously, getting the maximum number of fibers into a duct is a good idea. Reference 1 also comments that a cable diameter of 1 in (2.54 cm) fills 64 percent of a 31.8-mm duct. See the other rules related to this subject elsewhere in this chapter.

Reference

1. J. Lail and E. Logan, "Maximizing Fiber Count in 31.8-mm (1.25-inch) Duct Applications—Defining the Limits," *Proceedings of the National Fiber Optic Engineers Conference*, 2000.

PASSIVE OPTICAL NETWORK (PON) COST

An industry rule of thumb is $0.10/fiber/meter in a fiber cable (for the larger cables).

Discussion

Costs drive everything. PONs share an optical transceiver across a set of subscribers by use of a passive optical splitter. This allows multiple users to share the transceiver and fiber without active electronics or optics, such as optical amplifiers.

References

1. B. Lund, "Fiber-to-the-Home Network Architecture: A Comparison of PON and Point-to-Point Optical Access Networks," *Proceedings of the National Fiber Optic Engineers Conference*, 2001.
2. D. Gall and M. Shapiro, "Fiber-to-the-Home—Why Not Now?" Broadband Markets, at http://www.broadbandmarkets.com/articles/FiberHome2.htm.

2

Economic Considerations

An important component of any discussion related to telecom is its economics. The economics of bandwidth is integral to the design and development process, from networks to the smallest component. Cost/benefit analysis has driven the evolution of telecom. Telecom economics has allowed the deployment of the largest and fastest information system ever known to man, and this has already transformed our world, fundamentally touched our lives, and raised the human condition. The authors of this book made a concerted effort to include only economic models, rules, and data that reflect the conditions of the post-2000 economic meltdown.

Telecom networks are deployed to make money. A new network will not be deployed without sound economic forecasts showing that it will generate a profit. Witness the unfortunately slow adoption of the Internet in Africa, even though there is a suitable fiber backbone on that continent. It takes thousands of people and billions of dollars to deploy and support a nationwide network. Thus, it is critical for all who work in telecom to understand some of its basic economic tenets. This chapter contains a somewhat eclectic collection of rules, all relating to cost structures and economics inherent to modern optical telecom. Some of these are famous, and some apply to areas other than just telecom (such as the learning curve and Moore's law), but they all contain useful information for anyone engaged in any part of the optical telecom industry.

The telecom industry has developed an economic architecture based on several layers of companies as detailed by Gasmann[1] in the adoption (and modification) of his chart, given below. This chart mirrors the hierarchy of many industries wherein companies specialize in what they do best and provide those products and services to others with broader reaches. Conversely, large system integrators with broad reach and a systems perspective generally are not effective at component development. The consumer generally interacts only with the top of the chain and material suppliers at the bottom. The economic level is not a well defined boundary, and generally, a company will have products on more than one level. Often, new management will deliberately try to take the company to another level where high profits are smelled.

Economic level	Description of companies active at this level	Corporate example
Customer	Customers who subscribe to the end users for data and voice services.	John Q. Public
End users	Service providers and large network users who are buying and installing optical network gear.	AOL AT&T
Equipment system vendors	Manufacturers and/or marketers of equipment that can be used directly to construct networks that are bought by end users.	CISCO ONI, Nortel Ciena
Subsystem vendors	Manufactures and/or marketers of software-controlled products used to add complete functions to optical systems. Subsystem vendors sell primarily to equipment/systems vendors but occasionally sell to service providers.	Fujitsu Corning Novell Flextronics
Module vendors	Manufactures and markets multifunction boxes combining various components with easy interconnection. Module vendors sell primarily to equipment/systems vendors but may occasionally sell to service providers.	Flextronics EXFO JDS
Packaged component vendors	Companies that add connectorization and usable packages to components. These companies sell to equipment, subsystem, and module vendors.	Flextronics JDS Triquent
Component fabricators	The companies that manufacture optical semiconductors, lasers, detectors, and other components. These companies sell to equipment, subsystems, module and packaged component vendors.	Coherent Sensors Unlimited CVI ADI Intel Xilinx
Material suppliers	Supplies component vendors with materials such as silicon, InGaAs, GaAs, and InP.	Sensors Unlimited CREE Fermionics

The telecom industry isn't quite the free market some would lead us to believe. Frequency spectrum is controlled and sold by governments, and there are large government programs to encourage telecom networks. Governments encourage developing telecom networks in a number of ways. For instance, would the telephone have become so ubiquitous so quickly without the rural electrification program of the "New Deal" era? Rural electrification provided the power and telephone poles (for cable deployment), greatly reducing the cost of a telephone to a subscriber and network development for the phone companies. Local governments grant easements and special considerations to telecom assets. The Internet was invented by the U.S. Department of Defense's Defense Advanced Research Projects Agency in the 1960s as a communication network for large computers in the defense industrial complex and colleges. Parts of the backbone are still supported by the U.S. government via tax dollars.

The telecom industry went though an incredible economic expansion in the 1990s, with market capitalization soaring—often with little or no profits to justify it. The soaring stock market and corporate hiring were bolstered by forecasts of 70 percent per year growth rates ad infinitum. Old timers realize that nothing grows at 70 percent per year for long, but everyone seemed to think the bandwidth demand would, and that, more importantly, people would pay for it. The bubble burst in the spring of 2000, and it seemed to many that, once again, all the gold had been mined out of California.

Arden[2] points out that worldwide bandwidth demand is still being forecast at high annual growth rates. Even during the telecom bust, bandwidth seems to have continued to grow, although the growth rates for new systems have been down drastically due to excessive installed capacity. Carriers have been able to meet bandwidth demand by retrofitting networks to expand capacity. Mining this latent bandwidth in legacy hardware has proven to be a lower-cost and more efficient way of meeting increasing bandwidth demand. This trend will continue until carriers are forced to deploy new generations of system hardware, at which time equipment markets will rebound.

Given the economic upheaval in the telecom industry at the turn of the century, the authors of this book will refrain from suggesting any sources written before 2001. During the 1990s, it seemed to most observers that the economics of the telecom were changing the basic fabric of national and global economics. The interested reader would be served by going to basic economics books, as the business cycle doesn't seem to have been killed after all. Also, the references for the individual rules with a post-2000 date are worth reviewing.

In this chapter, the reader will find a collection of rules that relate to telecommunication economics. There are several rules about the cost of items such as a bits, photons, and lasers. There are also rules about the cost of deploying fibers in the ground and above ground.

Perhaps the most useful application of these rules is for detection of relative changes. This is especially true of the rules relating to the price elasticity of telecom attributes. Basically, an item with a large elasticity exhibits a market volume strongly in proportion to its price. The higher the cost, the smaller the market. These are generally luxury items such as vacation travel and recreational activities. Conversely, items with a small price elasticity exhibit little market change with price. These are generally necessities such as food and heating.

A seminal rule for the industry, called the "Learning Curve," is presented in this chapter. This economic stalwart has made and broken many corporations whose pricing was based on it. Originally developed about a century ago to estimate factory worker's performance, it seems to be a fundamental human economic property that can be applied to almost any activity. The key to success is to apply it correctly with the right percentage learning curve and to accurately estimate the number of units. Both are surprisingly easy to do with hindsight and surprisingly difficult with foresight.

Unfortunately, we did not find many rules that do not deal with capital outlays. Nor did we find many rules that deal with software, maintenance, or power consumption, although all are important considerations. It would be useful to include some rules concerning these issues, as well as QOS (quality of service) and revenue generation by differentiated wavelength services. Hopefully, readers will provide such for future editions.

References

1. L. Gasmann, "A Customer Oriented Approach to Optical Networking," *Proceedings of the National Fiber Optic Engineers Conference*, p. 1618, 2001.
2. Private communications with M. Arden, 2002.

COST IMPROVEMENT IN OPTICAL TELECOM

Historically, the number of bits per second per dollar of throughput has doubled every nine months for optical telecom networks.

Discussion

Many related technologies benefit from investment dollars and economies of scale, and we see their performance improve every year or so. Another famous rule in this chapter is Moore's law, which states that the capacity of an IC will double every 18 months. Anecdotal data from the late 1990s suggest that the number of bits that can be transmitted through an optical telecom network will double approximately every 9 months.[1]

As Strix states, "Photonics is at a stage that electronics experienced 30 years ago—with the development and integration of component parts into larger systems and subsystems. A rising tide of venture capital has emerged to support these endeavors. In the first nine months of 2000, venture funding for optical networking totaled $3.4 billion, compared with $1.5 billion for all of 1999, although this paced may have slowed in recent months. Investment in optical communication already yields payoffs, if fiber optics is matched against conventional electronics. The cost of transmitting a bit of information optically halves every 9 months as against 18 months to achieve the same cost reduction for an integrated circuit."

Reference

1. G. Strix, "The Triumph of the Light," *Scientific American*, Jan. 2001, pp. 81–86.

COST OF A BIT

1. Price varies as the square root of bandwidth.[1]
2. Long-haul (LH) installation costs are currently between \$4 and \$10 per gigabit per km.[2,3]
3. (Bits/s/wavelength) (no. of wavelengths) (km) $\propto$ corporate market capitalization.[3]

Discussion

Bitonomics dictates that the cost of transferring a bit from one point to another must decrease with time and the amount of bits transferred. RHK, Inc., a telecom industry analysis firm, estimated that the cost dropped an order of magnitude between 1993 and 1998, and other sources indicate a similar decrease in the cost of the communication of a bit. Historically, the cost per bit dropped quickly as newer, high-capacity systems were deployed. As Cooperson[4] explains, "As the throughput of transport systems has increased from 2.5 Gb/s over one channel in 1994 to over 400 Gb/s over multiple 2.5 or 10 Gb/s channels, the cost of transport in thousands of dollars per Gb/s has decreased from over \$200k to under \$20k—an order of magnitude reduction in five years."[4] With the economic doldrums of the turn of the century, the rate at which the cost to transport a bit is not falling as fast as it has historically but, obviously, over the long term, the costs should continue to fall.

The above costs are for the equipment to light up a bit and do not include deploying the fiber cable (fiber costs are addressed by other rules in this chapter). The cost to deploy a long-haul system to move bits from one point to another is about \$4 to \$10 per gigabit per km as of 2002, and somewhat lower for the rare ultra-long-haul (ULH) networks (e.g., \$3 to \$4 per gigabit per km). As time passes, costs are expected to decrease, driven by dense wavelength division multiplexing and integration of all optical components. This will lead to costs well under \$1 per gigabit-mile as explained in the "Cost of Deploying a Fiber Cable Underground" rule. This falling price per bit is especially true for equipment pricing and, incidentally, seems to hold for costs to the consumer as well.

The third bullet above relates bit rate to a company's market capitalization (the price of the stock multiplied by the number of outstanding shares). During the wild 1990s, this seemed to hold true. A network company could double its bit rate or double its range and see its stock double. However, Wagner points out that this is no longer true. "Sending bits faster and further in a straight line was sufficient in an industry focused on transmission. But building networks, instead of links, requires more intelligence and added value. Unlike physics-driven devices with little software content, emerging components and subsystems will emphasize flexibility and reconfigurability and include more software than their predecessors."[3]

Figure 2.1 indicates the cost to transmit a gigabit for a kilometer plotted by the length in kilometers for LH and ULH. Arden[2] conducted a detailed theoretical cost model to generate these curves based on theoretical pricing points and the development of a benchmark network based on characteristics of deployed networks in 2000, assuming high efficiencies. In reality, the costs are likely to be higher, due to real-world inefficiencies in the architecture.

The average price per gigabit was determined by adding up the cost of subsystems (including terminals, regenerators, amplifiers, transmit-receive devices, and optical add/drop modules) needed at a predetermined distance interval plus other associated costs (e.g., the cost of initial equipment deployment, dispersion compensation, software, etc.). These other costs were determined, based on analysis in KMI's dense wavelength division multiplexing (DWDM) report, to be about 20 percent, on average, of the total cost for a single systems deployment. Additionally, about 40 percent of the total system price was for margins on components. This number is factored into the model to provide an example of what the price of a system would be and not the cost of producing the system. This final capital expenditure figure was then divided by the average capacity determined from the weighted averages, noted earlier, at the predetermined distance interval. The average price per gigabit was then divided by the maximum distance number in the predetermined distance interval (intervals were done at 50 km for distances up to 600 km and 100 km at distances above 600 km). This determined the price per gigabit per kilometer at the specified distance interval.[2]

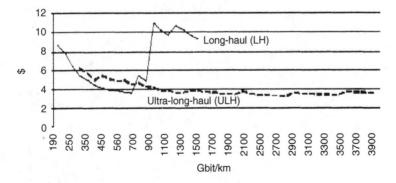

Figure 2.1   Dollars per gigabit per kilometer.

This model was applied to the LH arena and then applied to the ULH equipment arena. The results were then placed onto a line graph for visual comparison.

The LH has the assumption that service providers would maximize capacity, while the ULH assumption is that service providers would design the most distance-efficient system. One can see that the ULH cost is relatively flat beyond distances of 1500 km or so, at just under $4/Gb/km. Again, these numbers include more than the fiber, so they are substantially higher than determined by the fiber-cost rules presented elsewhere in this chapter. Arden explains, "The intersection point of the long-haul solution and ULH solution is at the point where the first O-E-O regeneration is needed (at just under $4.50/Gb/km). The loss in capacity efficiency and the need for O-E-O regeneration in the long-haul solution ensures that the ULH solution remains more cost effective..."[2] for distances greater than 1000 km.

References

1. F. and R. Menendez, "Hybrid Copper Fiber Coaxial Access Network," *Proceedings of the National Fiber Optic Engineers Conference*, 1999.
2. M. Arden, "The Ulta-Long-Haul Market: How Big a Stretch Is It?" *Proceedings of the National Fiber Optic Engineers Conference*, pp. 1263–1264, 2001.
3. P. Wagner, "The Next Wave of Optical Networking: A Flight to Quality," *Proceedings of the National Fiber Optic Engineers Conference*, p. 197, 2001.
4. D. Cooperson, "The Evolution of DWDM to Optical Networking Will Develop in the Metro Market in 2000," *Proceedings of the National Fiber Optic Engineers Conference*, 1999.

COST OF DEPLOYING OPTICAL CABLE IN THE AIR

The cost to deploy an aerial fiber cable is approximately \$11,000 to \$15,000 per mile (or \$6800 to \$9300 per km) and can be approximated by

$$\$ \approx 13,000(L)$$

where $\$$ = cost to deploy a 24-fiber cable
   L = length of the deployment in miles

Discussion

Aerial deployment of fiber is generally less expensive than underground deployment, if the poles and right of way already exist. The comment about the "right of way" deserves additional comment. In high-rent neighborhoods, this can be the driving cost, but it is of little concern when crossing Bureau of Land Management (BLM) land in Nevada. However, in general, the rights of way for aerial cables have existed for a long time for telecommunication companies (often over a century in North America and Europe).

McNair[1] gives cost approximations for aerial deployment of three types of 24-fiber cable. They are summarized in the tables below. A *lashed cable* is a traditional method whereby a support cable is deployed, and the optical cable is "hung" from the supporting cable. "Figure-8" cables are ones in which the top part of the cable is a mechanical support, and the lower part is the optical portion; they generally have a figure-8 cross section. *All-dielectric self-supporting (ADSS)* cables are generally based on loose tube structures with integrated tensile supporting members built into the single cable structure.

As McNair explains,

> Field information has also suggested that the typical cost for a contractor to place a messenger wire, then install and lash an optical cable, is estimated to be approximately \$2.0/ft. This cost represents a constant loaded labor/equipment rate and includes provision for overhead expenses, installation equipment, depreciation, etc. Since each of the installation methods will incorporate essentially equivalent costs for additional pole guying and splicing, these additional costs have not been considered in the following analysis. Also, additional hardware costs for such items as closures and slack storage devices, etc. have not been included. Similarly, equipment and labor costs to splice the cables together have not been included.[1]

The cable is assumed to be a 24-fiber cable placed on poles 300 ft apart.

Type	Installation cost in dollars	Days to complete 10,000 ft run	Cable price	Messenger	Mounting hardware	Total cost
Lashed	20	2	10.1	0	1.7	21.8
Figure-8	13.3	1.33	8.3	0.9	0.7	29.8
ADSS	10	1	13.7	0	0.3	27.4

One can see that, for the sake of a "rule of thumb," all total costs are similar, with the ADSS cables being the least expensive to install. The old method of lashing the cables is the most expensive approach. If one picks a middle range and does a simple $Y = mx$ curve fit, the equation on the previous page can be generated.

As obvious caveat is that the validity of these relationships is based on "average installations," and it can be affected items such as

- Difficult terrain
- Crew experience
- How far the deployment is from a road
- New cable vs. existing cable
- Even the weather

These may change the time and cost required. Also, these estimates are for deployment and do not include cable maintenance costs, which are lower for underground deployment. Aerial deployments encounter constant maintenance threats from trees, icing, and creep in addition to the same problems with animals and inadvertent breakage sustained by underground cables (although the latter two occur less frequently in aerial cables). The above also does not include the equipment costs to "light" the cable, which is several times the cost of deploying the fiber cable.

Reference

1. J. McNair, "Installation Cost Scenarios...Self-Support, Lash and Overlash Cables," *Proceedings of the National Fiber Optic Engineers Conference*, 2000.

COST OF DEPLOYING A FIBER CABLE UNDERGROUND

It costs about $70,000 per mile to install fiber cable in the ground.

Discussion

Although raw fiber costs about $100 per kilometer, the cost of fiber is usually inconsequential when compared with actually deploying a cable and almost negligible when compared to the cost to actually install and activate a link, which tends to be a few dollars per gigabit per kilometer.

The process of assuring right of way, marking buried utilities, closing streets, installing ducts, digging up the ground, and installing and initially testing the cable costs between $50,000 and $70,000 per mile, including the hardware. As noted by Redifer,

> Industry standards range from approximately $35,000/mile in rural areas to approximately $100,000/mile in urban areas. Using an average of $50,000/mile, a 600-km long-haul span would cost approximately $18,750,000 to plow in. Both the wavelength division multiplexing (WDM) and time division multiplexing (TDM) solutions are more cost effective than installing new fiber when increasing capacity on existing fiber routes.[4]

An obvious caveat is that this applies to an "average installation." Such items as difficult terrain, crew experience, new cable vs. existing cable, and even the weather may change the time and cost required.

Underground cable costs more than aerial cable but has substantial advantages in terms of neighborhood appearance, and maintenance. Although underground cable may be more frequently cut by construction (as it is out of sight), repair costs are inconsequential as compared to aerial line hazards such as ice storms, tornados, hurricanes, falling trees, and auto accidents.

The above costs are for a cable, which may contain many fibers. Thus, the cost per fiber is the above divided by the number of fibers that the cable contains. If the cable contains 72 fibers, then the cost is more like $700 to $1000 per mile per fiber. Additionally, if each fiber contains 200 laser lines, then the cost per telecom link is further reduced by a factor or 200. If each of those laser lines has a 40-Gb data transmission speed, then the cost is merely 9 to 13 cents per Gb-mile, compared to dollars per mile for legacy lower-rate, fewer-lambda systems. Note that these cost calculations do not include the cost of repeaters, amplifiers, or add/drop and other network elements, all of which can push the cost substantially higher, as explained in other rules in this chapter. This is the cost of the fiber, not the network.

Since most of the cost of deploying a fiber comes from labor and incidentals (rights of way, road closures, boring, and so on) as opposed to the cable itself, it is generally wise to deploy as many fibers in a cable as possible. As Stevens points out,

Installing the proper cables now avoids revisiting these costs at the next upgrade. When cost modeling is reviewed later, it will be seen that the cost of fiber optic cable is typically only 8 to 12 percent of total installed cost for an optical access network. Given these parameters, the incremental cost of fiber is approximately $0.06 to $0.12 per fiber meter, depending on cable fiber count. Two thousand feet of 72-fiber cable costs about 14 percent more than that of 60-fiber cable, which amounts to several hundred dollars. This incremental cost is easily justified when considering the emergence of future services and the difficulty of adding a completely new 12-fiber cable.[5]

Finally, to add the same bandwidth by installing dense wavelength division multiplexing (DWDM) on the same cable involves approximately onesixth of the cost of deploying a new fiber cable with the same capability, assuming the fiber is deployed.

References

1. S. Shepard, *Optical Crash Course*, McGraw-Hill, New York, p. 126, 2001.
2. A. Girard et al., *Guide to WDM Technology and Testing*, EXFO Electro-Optical Engineering Inc., Quebec City, Canada, p. 11, 2000.
3. S. Greenholtz, "Deployment of Metropolitan WDM Ring Technology in the Bell Atlantic Network," *Proceedings of the National Fiber Optic Engineers Conference*, 1999.
4. G. Redifer, "DWDM vs. TDM in Metro Long-Haul Applications," *Proceedings of the National Fiber Optic Engineers Conference*, 1997.
5. S. Stevens, "Migration to All Optical Network," *Proceedings of the National Fiber Optic Engineers Conference*, pp. 688–689, 2001.

COST OF LASER TRANSMITTERS

- Tunable lasers cost approximately $2500 and are expected to cost about half that in a few years.
- Tunable lasers result in great cost savings for DWDM networks.
- Fixed wavelength lasers cost about $1000 and are expected to cost about half that in a few years.
- The electronic line cards to drive lasers cost about $25,000.

Discussion

These costs represent general ranges, experienced in 2002, and the projections are for the 2004 to 2007 time frame. Obviously, any cost projections can turn out to be very wrong with simple changes in the industry, manufacturing process improvements, or technical breakthroughs.

The telecom industry has always been cost driven, but the desire to be cost effective has intensified since the 2000 stock slump. More expensive components have to be justified on a cost basis as well as a technical basis. Tunable lasers have significant cost benefits for DWDM networks, including reducing the number of spares[*] as well as reducing the floor space required for equipment and providing greater growth flexibility.

Thomas and Chang point out that

> Because they allow significant simplification in intelligent network design, which corresponds to cost savings, tunable lasers need not match the price of fixed lasers exactly. Rather, a premium higher than that of that of a fixed laser product could be sufficiently competitive. Another hurdle is the need to pass stringent qualification tests for long-term stability. A third may be the capability of plugging into the same circuit as a fixed wavelength lasers. So far, it looks like all of these hurdles are surmountable.[1]

For WDM systems with just a handful of laser lines, provisioning spares with fixed-wavelength lasers was costly but affordable. For DWDM systems of 80 or more laser lines, the amount of hardware can be prohibitively expensive to purchase, maintain, and store. The following three quotes illustrate the cost benefits of tunable lasers for DWDM systems:

> Consider the sparing problem at a central office (CO) with a large number of operating fixed-wavelength lasers. A fixed wavelength approach requires a local inventory for each wavelength of each card that is ITU grid specific. The cost and space required for the sparing

[*]For example, with individual lasers, you have to have spares at each wavelength; not so with tunables, so the system cost associated with spares drops dramatically.

is massive. For example, for a simple point-to-point system carrying 80 wavelengths, with each of the line cards at a cost $25,000, the sparing cost is $2 million, not including G&A costs. A couple of line cards, equipped with tunable wavelength lasers, can eliminate the need for all spare line cards in the entire C- or L-band and provide the same redundancy and reliability for a fraction of the investment.[1]

A large number of spare lasers

...also means a large number of transmitter boards containing other essential components such as laser drivers, modulators and driver ICs. For each wavelength channel operating at 10 Gb/s, this could cost as much as $50,000. Instead, using a tunable laser, one board can be held as backup and tuned to the wavelength of the faulty fixed wavelength laser. These same boards can then migrate to replace all fixed wavelength lasers in the network, as their cost approaches that of a fixed laser.[2]

With a desire for

...20 lasers for each wavelength, a 160-channel system would require 3200 fixed-wavelength lasers for inventory. At a price of $1000, the cost would be $3.2 million for laser inventory alone. Although tunable lasers are more expensive than their fixed-wavelength counterparts, placing 20 lasers in inventory to cover the same requirements of 3,200 lasers has obvious cost benefits for the network operator.[3]

References

1. M. Thomas and Y. Chang, "System Advantages of Tunable Lasers in the Metro DWDM Environment," *Proceedings of the National Fiber Optic Engineers Conference,* pp. 1487–1490, 2001.
2. http://keyconference.com/plastow/plastow.htm.
3. http://www.techmall.com/techdocs/TS010910-6.html.
4. http://www.agility.com/what/3040.cfm.
5. http://zdnet.com.com/2100-11-528957.html?legacy=zdnn.
6. http://buyalternatives.tm.agilent.com/specials.shtml.

Cost of a Photon

A photon from a bulb costs about $\$10^{-25}$; a diode laser photon runs about $\$10^{-22}$.

Discussion

The above is based on the current costs to buy and operate a light bulb and semiconductor laser.

For example, a 60-W household light bulb costs about $2 and consumes about 60 kWh of electricity in its approximately 1000-hr lifetime. It has a tungsten filament that is raised to a temperature between 2000 and 2900 kelvins (K). If we assume that the tungsten filament's temperature is 2600 K and has a square centimeter of total area (with an emissivity of 0.45, which is typical for hot W), it will emit about 2.2×10^{19} photons per second in the visible bandpass. Infrared and UV photons that are emitted from the filament don't count, as they do not make it through the glass bulb but, instead, rattle around inside and heat up the bulb structure. One thousand hours is 3.6 million seconds, so the bulb will emit about 7.8×10^{25} usable photons in its lifetime, at a cost of about $10. Thus, the cost of a bulb-generated photon is about 10^{-25} dollars.

Additionally, a 10-mW diode laser can be purchased for $10. It operates for approximately 1 million hours before it fails, but it emits all of its specified light in its given narrow bandpass. At 1.5 microns, a 1-mW laser emits 7.5×10^{16} photons per second and 7.5×10^{22} photons over its lifetime.

A telecom transmitter has other costs added to it, such as certification, warranty, and modulation mechanisms. Generally, these produce more than 10 mW and cost much more than $10. The cost of a telecom photon is more like 10^{-20} dollars.

Reference

1. P. Hobbs, *Building Electro-Optical Systems, Making It All Work,* John Wiley & Sons, New York, pp. 64, 68, 2000.

DECLINING COSTS

- Integrated circuit costs decline approximately 25 to 35 percent per year.
- Optics decline in price by about 10 to 20 percent per year.

Discussion

Generally, one can assume that the costs of integrated circuits and optics are declining. This results from improved yields and amortization of capital equipment and nonrecurring engineering. The amount that they decline each year varies as a result of a myriad of variables such as demand, material costs, labor rates, and so on. The above are frequently cited rates of declining costs used in cost models.

Greater declines are generally realized when buying a standardized piece of optics or IC or module containing them. Even a slight modification (such as a coating) will often result in the elimination of any price decline.

Much of this decline has to do with economies of scale resulting from increased marketing success, more users, and more applications. Other reasons include the "learning curve" described elsewhere in this chapter. Additionally, the IC cost reduction can be also attributed to larger wafers that result in more chips, and purer wafers as described in the chip yield rule in Chap. 3.

The above can be put into mathematical form as the following equation, assuming that costs and price track linearly (which they tend to in the long run):

$$C_y = \frac{C_i}{e^{Ay}}$$

where C_y = cost at year y
C_i = initial cost (at year zero, or the present)
e = mathematical constant, 2.7183....
A = a constant, per the above rules and equation; this works out to be about 0.36 for generic integrated circuit costs and approximately 0.16 for optics
y = year in question for the calculation

Reference

1. J. George, et al., "Laser Optimized Multimode Fibers for Short Reach 10 Gbps Systems," *Proceedings of the National Fiber Optic Engineers Conference*, p. 353, 2001.

COST OF AN OUTAGE

Outages can result in lost revenue of $77,000 per hour and take an average of about ten hours to repair.

Discussion

A single OC-192 channel can carry 128,000 separate phone calls. If the provider is allocated $0.01/min for these calls, then the fiber link is capable of generating $1280/min or about $77,000/hr. The average mean time to repair (MTTR) a broken link was 9.7 hours according to one survey.[1]

Obviously, an actual link has a mix of data and telephony traffic. Unfortunately, it is harder to put a price on data traffic, as the outage tends to be more of a nuisance for the user, who doesn't get to send his manuscript to his co-author when convenient and instead must log on again and try to do it some other time. However, there is a quality of service (QOS) issue with frequent outages, as they will drive a user to another service, which results in lost revenue. Also, the above assumes a full-capacity OC-192 with the telephone connections; generally most links are not operating at full capacity. Conversely, it is possible that some large trunk lines could result in even more lost revenue if the provider was receiving more than $0.1/min/line or had fines imposed by the government for outages.

The causes of outages vary from equipment failures in the stations to cable breaks. There is an associated rule in Chap. 7, called "Outage Rates," in which the percentage of outages for such diverse causes as rodents and gunshots is given.

The economic considerations of outages are interwoven with systems for automatic monitoring (such as optical performance monitors) and fiber break monitors such as optical time-delay reflectometry (OTDR), and there are rules describing attributes of these in Chap. 6. Such early-warning devices (before the irritated user calls) can be built into the network as a remote fiber test system (RFTS). Coker details the cost benefits of such a system:

> For a given network, the cost of an RFTS can be estimated from the following guidelines:
>
> - Cost per monitored site is approximately twice that of a portable OTDR.
> - Number of monitoring sites is comparable to the number of portable OTDRs purchased to maintain the network.
> - The RFTS equipment has a minimum lifetime more than twice that of today's OTDRs, which leads to an amortized cost comparable to portable field equipment.[2]

Coker continues,

Surveying current GN Nettest RFTS network monitoring customers, we find that mean fault isolation and localization times were approximately 2–5 minutes with network monitoring and 2–48 hours
without network monitoring, depending on the particulars of the
fault. Dispatch, repair, and turn up times were comparable with or
without a monitoring system. In terms of average revenue per cable,
there was a large range of values dependent on fiber count and provided services. Penalties of course are a function of operating coun-
try and SLA contracts.[2]

References

1. S. Chabot and G. Levesque, "DWDM Impact on Traditional RFTS: Monitoring
   Colors," *Proceedings of the National Fiber Optic Engineers Conference*, 2000.
2. D. Coker, "New Business Case for Integrated Network Monitoring," *Proceedings
   of the National Fiber Optic Engineers Conference*, 2000.
3. J. Chamberlain and D. Vokey, "Metallic Armored vs. All Dielectric Fiber Optic
   Cable, the Pros and Cons," *Proceedings of the National Fiber Optic Engineers Conference*, 1998.

THE LEARNING CURVE

Each time a succeeding unit is made, it takes less time (and less cost) to build than did its predecessor. The time can be estimated by a decrease every time the number of units made is increased by a factor of two.

$$C_n = C_1 N^{([\log PLR]/0.3)}$$

where C_n = time it takes to make the nth unit in production
C_1 = time it takes to make the first unit
N = production number of the unit
PLR = percent learning rate in decimal notation

Discussion

This rule was initially based on empirical studies of production lines. It has been applied successfully to all kinds of activities. Although the development of learning curves was based on and calibrated by examining touch labor, they can be applied to any variable cost, such as material, and sometimes even to some traditionally fixed costs such as sustaining engineering or management. Therefore, they can be used loosely to estimate total cost and total price directly by using the first unit price in the above equation.

Learning curves are based on the fact that every time you increase the number of units that you make, you will experience a predictable decrease in the time that it takes you to make the next unit. This production phenomenon, in which each succeeding unit is easier to make, really occurs. The learning curves provide a powerful tool to estimate the reduction in costs or time for a production run. Surprisingly, they can be amazingly accurate. The trick is to know what PLR to use.

The 0.3 in the equation is $\log_{10} 2$, because PLR is defined as the improvement experienced when production doubles. If the first unit of a production sensor takes 10 hr to test and the second takes 9 hr, the PLR is 90 percent, and 0.9 should be used in the equation. Typically, optical telecom components usually have learning curves in the eighties, and simple mechanical assemblies or integrated circuit yields may get into the seventies.

Again, the biggest caution is to be sure that you are applying the correct PLR, as costs for large-scale production are very sensitive to this parameter. The above equation assumes that the learning rate is constant; in fact, they usually are not. Typically, there is a lower (better) PLR in the beginning as manufacturing procedures are refined and obvious and easy corrections to the line are applied. A very mature line may actually experience the opposite effect, whereby it takes longer to make the next unit because of tooling wear and because old capital equipment keeps breaking.

Let us state that you develop a product and that it cost $10,000 to produce the first prototype after several million dollars of nonrecurring research, development, and engineering expenses (for a production system, these "one-time" costs are to be excluded). Let's also assume that this electro-optical product follows typical learning curves for optical system manufacturing of ~90 percent and that you are conservative and expect the next unit (the first one that you charge to the customer) to cost as much as the prototype. A quote for five of these is immediately requested. From the above equation, you can develop the chart below and assume that it will cost you a little over $43,000 to produce the five units. If your company requires a 50 percent margin, you double the cost for the price and tell the perspective customer that a price for five would be approximately $86,000.

Production unit no.	Cost per unit, $*	Total cost, $†	Average unit production cost, $
1	10,000	10,000	10,000
2	9,000	19,000	9,500
3	8,462	27,462	9,154
4	8,100	35,562	8,890
5	7,829	43,391	8,678

*Use the equation in the rule to obtain the cost of each successive unit.
†Each is added to the previous column.

Note that the equation provides the cost for the nth unit. For the entire manufacturing run, you must integrate (or add up) the cost of all of the units. Additionally, for an average unit cost, you must divide the total cost by the number of units. This can be done easily via numerical integration with a spreadsheet or math program.

Now let us say the president of your company calls you in to inquire about the network prospects of your hardware. He wants to know what the 1000th unit would cost to make. You get aggressive with the learning curve for such a large number (assuming that the company will invest in automated production facilities), apply an 80 percent PLR, and come up with the following projection:

$$C_n = C_1 N^{([\log PLR]/0.3)}$$

$$C_{1000} = (10,000)(1000)^{([\log 0.8]/0.3)} = (10,000) \times (1000^{-0.323})$$

and come up with an estimate for the cost of the 1000th unit of $1100.

LINK AND NODE COST IN A WAVELENGTH ROUTED OPTICAL NETWORK

The cost of each link (i) is the summation of the cost of active (x_i) λ-channels and the cost of adding a new wavelength channel on link i.

$$C_i(x_i + 1) = C_i(x_i) + \Delta_i$$

where

$$\Delta_I = \left\lfloor \frac{L_{Link(i)}}{L_{Span} \times [\#Span]_{(i)}} \right\rfloor \times C_{OTU} + f(Mod(x_i/W_i))$$

$$\times \left[L_{Link(i)} \times C_{Fiber} + \left\lfloor \frac{L_{Link(i)}}{L_{Span}} \right\rfloor \times C_{OA} + \left\lfloor \frac{L_{Link(i)}}{L_{Span} \times [\#Span]_{(i)}} \right\rfloor \times C_{Reg} + 2 \times C_{ET} \right]$$

$$+ f(Mod(x_i/(W_i \times F_i)) \times \infty)$$

$$f(Mod(x_i/W_i)) = \begin{cases} 1 \text{ If } Mod(x_i/W_i) = 0 \\ 0 \text{ If } Mod(x_i/W_i) \neq 0 \end{cases}$$

$$f(Mod(x_i/W_i \times F_i)) = \begin{cases} 1 \text{ If } Mod(x_i/W_i \times F_i) = 0 \\ 0 \text{ If } Mod(x_i/W_i \times F_i) \neq 0 \end{cases}$$

where $C_i(x_i)$ = link cost for active λ–channels on link i
$C_i(x_i + 1)$ = link cost for active λ–channels and a new λ–channel on link i
Δ_i = cost of adding a new λ–channel on link i
W_i = maximum number of wavelengths per fiber pair on link i, $W \geq 2$
F_i = maximum available fiber pairs on link i
x_i = number of active λ–channels on link i
$x_i + 1$ = number of active λ–channels plus a new λ–channel
L_i = length of link i
L_{Span} = transmission span distance between two adjacent optical amplifiers
$\#Span$ = number of spans before signal regeneration
C_{OTU} = cost of one OTU
C_{Fiber} = fiber cost per unit distance

C_{OA} = cost of one optical amplifier
C_{Reg} = cost of one regenerator
C_{ET} = cost of one end terminal

The cost of node j is the summation of the cost of active y_j λ-channels plus the cost of adding a new wavelength channel in mode j.

$$C_j(y_j + 1) \;=\; C_j(y_j) + \Delta_j$$

$$\Delta_j \;=\; 1 \times C_{Port} + f(Mod(y_j/CH_{j(OXC)})) \times C_{OXC}$$

$$f(Mod(y_j/CH_{j(OXC)})) = \left\{ \begin{bmatrix} 1 & \text{If } Mod(y_j/CH_{j(OXC)}) = 0 \\ 0 & \text{If } Mod(y_j/CH_{j(OXC)}) \neq 0 \end{bmatrix} \right.$$

where $C_j(y_j)$ = node cost for active λ-channels in node j

$C_j(y_j + 1)$ = node cost for active λ–channels and a new λ–channel in node j

Δ_j = cost of adding a new λ–channel in node j

$CH_{j(OXC)}$ = maximum switch size per optical cross connect in node j

C_{Port} = cost of one pair of directional ports

C_{OXC} = cost of each OXC

Discussion

A network's cost can be simplified to the summation of the cost of all of the links and all of the nodes as follows:

$$C_{total} \;=\; \sum_{i=1}^{L} C_{i(link)} + \sum_{j=1}^{N} C_{f(node)}$$

The routing optimization problem is to provide the lowest cost network that can be achieved by minimizing the above C_{total}. The source suggests a heuristic algorithm using a Monte Carlo approach to mitigate the problems of stopping at local minima and not the global minima.

The above equations indicate how to determine the equipment cost of the links and nodes for a wavelength routed optical network. Note that the equations give a relatively simple accounting of all of the component costs, many of which can be estimated from other rules in this chapter.

Hu and Mezhoudi[1] point out their assumptions for the pricing mechanisms as listed below:

The network total cost increases with each new wavelength channel; however, the increase is not linear with each additional channel in realistic network building up, since it is much more expensive to activate a λ–channel in a new fiber than in an existing fiber, and it is also expensive to activate a λ–channel in a new network equipment than in an existing network equipment. Based on above factors, we suggest the following incremental pricing mechanism:

- When the number of active λ–channels in a fiber is more than 0 and less than fiber maximum capacity (W), a new λ–channel activation does not require setting up a new fiber, and the cost increase is the amount of *optical translate units (OTUs)* for this new λ–channel.
- When the number of active λ–channels modulo W is 0, setting up a new λ–channel would require a new fiber pair or a new bidirectional fiber and corresponding network equipment such as optical amplifiers, regenerators, end terminals, and OTUs.
- When the number of active λ–channels switching in an optical cross-connect is more than 0 and less than the optical cross-connect maximum switching size (CH_{OXC}), a new λ–channel activation does not require setting up a new optical cross-connect, and the cost increase is the amount of OTUs for this new λ–channel.
- When the number of active λ–channels modulo CH_{OXC} is 0, setting up a new λ–channel would require a new optical cross-connect.
- When the number of fiber pairs is more than 0 and less than maximum number of fiber pairs on link (F_{max}), a new λ–channel activation is permitted.
- When the number of active λ–channels modulo W^*F is 0, a new λ–channel activation is not permitted.

Reference

1. Y. Hu and M. Mezhoudi, "Cost Effectively Design an Optical Network Based on Optical Signal Noise Ratio (OSNR) Requirement," *Proceedings of the National Fiber Optic Engineers Conference*, pp. 1037–1039, 2001.

MOORE'S LAW

- The number of transistors on a chip double every 18 months.

Often, this is paraphrased to read

- The power of a microprocessor (or memory on a chip) doubles every 18 months.

This also is often paraphrased to read

- The processing power per unit cost doubles every 18 months.

Discussion

This is perhaps the most ubiquitous "rule-of-thumb" in this tome. It contains all the aspects that the authors of this text like about any rule of thumb—it is simple, useful, and easy to remember and calculate with a hand calculator in a meeting; it sheds in sight and intuition and is surprisingly accurate.

In 1965, Gordon Moore (of Intel) extrapolated, from the capability of early chips, that the number of transistors on a chip would about double each year for the next ten years. He revised this law in 1975 to the doubling of the number of transistors every 18 months. According to the Intel web site,[1] "In 26 years, the number of transistors on a chip has increased more than 3,200 times, from 2,300 on the 4004 in 1971 to 7.5 million on the Pentium® II processor." As an interesting aside, Moore also estimates that more than 100 quadrillion transistors have been manufactured as of 2002.[2]

Even though the driving forces in this rule include micro- and macroeconomics, the development of capital equipment, chip technology, and even sociology, Moore's simple extrapolation has proven to be quite accurate.

The final paraphrase implies a simple mathematical equation that can relate cost or price drop every 18 months to a given (constant) processing power as

$$C_y = \frac{C_i}{e^{0.463y}}$$

where C_y = cost for a given amount of processing at year y
C_i = initial cost (at year zero)
e = mathematical constant 2.7183....
y = difference in years between i and y (initial and projected)

Recently, many individuals have cautioned that, due to the approaching quantum effects and diffractive effects with x-ray photolithography, Moore's law may slow and even stop when the feature size approaches about 10 nm. "Physicists tell us that Moore's law will end when the gates that control the flow of information inside a chip become as small as the wavelength of an electron (on the order of 10 nm in silicon), because transmitters then cease to transmit."[2] This feature size is bound to be reached on a commercial level sometime between 2010 and 2020.[2]

Additionally, Takahashi[3] astutely indicates that the barrier to Moore's law might well result not from miniaturization but from more mundane attributes such as testing, reliability, and the equipment failure rate when going to smaller-scale and newer techniques. He also points out that the delay in the development cycle can be large, and the risks get larger with each increase in density. He observed, "The risk of a six-month delay have grown from 1 to 3 percent a few years ago, with the 250-nm generation, to 15 to 30 percent for the 130-nm generation, according to the design tool maker Synopsys, based on a survey of its design teams."

However, these naysayers are selling human intuition short and potentially ignoring the human insatiable need and ability to pay for information and technology. First, new materials can extend Moore's law beyond what is predicted, albeit with a massive (trillion dollar) investment in new foundries. Second, quantum effects aren't all necessarily bad; they can be potentially tamed to extend the "effective feature size" even lower than the wavelength of an electron.

Moreover, silicon hasn't quite lost its edge, as three-dimensional or vertically integrated chips hold the promise of keeping Moore's law alive for another 20 years or more in silicon. This technology is being actively pursued by several companies, including Irvine Sensors, Matrix Semiconductor, Raytheon, and Ziptronix. Other techniques (e.g., IBM's silicon on insulator), novel interconnects, and innovative micropackaging may also allow the effects of Moore's law to continue for several decades.

Moore's law has held up surprisingly well for several decades, but we are clearly approaching a turning point. The future rate of expansion in the number of transistors per chip is questionable.

References

1. http://www.intel.com/intel/museum/25anniv/hof/moore.htm, 2001.
2. T. Lee, "Vertical Leap for Microchips," *Scientific American,* Jan. 2002, pp. 52–59.
3. D. Takahashi, "Obeying the Law," *Red Herring,* April 2002, pp. 46–47.
4. J. Westland, "The Growing Importance of Intangibles: Valuation of Knowledge Assets," *BusinessWeek CFO Forum,* 2001.

NETWORK COST

A quick comparison of network cost can be estimated by

$$
N_c = \frac{V_t TL}{2C_p}\left(\frac{\sqrt{N}-1}{\sqrt{N}+1}\right) + \left(\frac{0.4\,TLV_t}{NC_p\sqrt{N}}\right) + (ANB)\left(\frac{V_t\sqrt{N}}{C_p}\right)
$$

where N_c = network cost

   N = number of routers

   V_t = volumes of traffic in terms of port capacity

   C_p = port capacity

   T = cost of a mile of fiber

   L = linear size of the region served by the network

   A = fixed cost of a router

   B = fixed cost of a port

Discussion

Everyone always tries to minimize network cost, as it is the Holy Grail of a network company's profitability. However, this is not a trivial issue, because one can often minimize costs as a function of a single variable. Life becomes very murky when trying to find that elusive "universal minimum." One could develop a cost model as a function of performance, provisioning, maintainability, growth options, or any number of other parameters. These all must be carefully considered in the early stage of network architecture. The above equation provides a good start to estimate the network deployment costs as a function of the number of router systems and to trade that off with the fiber deployment costs (see Fig. 2.2).

The equation starts from the assumption that the location of the points-of-presence (POPs) are uniformly distributed so that the average distance between router and POP decreases as the number of routers increases. Also assumed is that the grid is square, so very odd shaped networks will be different. The reference's authors started by curve fitting empirical data to obtain the first relationship for the average distance to the router.

Vendors frequently give volume discounts to network deployers, and the above equation assumes a fixed cost. However, the user can modify this equation for significant volume discounts. Obviously, the cost of ports and router systems are composed of component and testing costs.

The developers of the equation (Babbitt and Zwall) astutely point out an interesting feature: In the voice, satellite, and wireless cases, the network cost

...exhibited a $1/\sqrt{N}$ relationship. The topology of the voice, satellite and wireless scenarios with the $1/\sqrt{N}$ cost relationship resulted

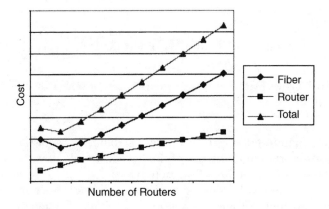

Figure 2.2   Number of routers vs. deployment cost.

in optimal network designs with fewer larger switches as the fiber cost decreased. This particular topology yielded a different result. From this we are forced to conclude that, in the near term, the cost of a nation wide IP backbone is minimized with a few large routers, regardless of backhaul costs or router costs. This will continue to be the case until routing costs fall to the point where a routing versus backhaul trade-off is meaningful.[1]

Remember (see the related rule, p. 39) that a mile of deployed fiber cable costs about $70,000, so T is equal to $70,000 divided by the number of fibers in the deployed cable.

Reference

1. J. Babbitt and J. Zwall, "Optical Router Size Analysis in a Nationwide IP Network," *Proceedings of the National Fiber Optic Engineers Conference*, 2001.

METCALFE'S LAW

Network value rises by the square of the number of terminals.

or

The value of a network grows by the square of the processing power of the terminals attached to it.[1]

Discussion

The economic engine behind network development is that the cost of the network tends to be proportional to the size of the network, while its value is proportional to the square of the size of the network.

A classic non-computer network example is voice telephones. A single telephone isn't worth anything, because there is no one to call. As stated by Theodore Vail (president of AT&T) in 1908, "A telephone without a connection at the other end of the line is...one of the most useless things in the world. Its value depends on the connection with other telephones— and increases with the number of connections."[2]

By the 1920s or early 1930s, when most businesses and some people owned phones, the device started to have value. By the 1960s, almost every household in developed counties had at least one phone, and most businesses had several—some thousands. The value of telephones did not stagnate; the power of the phone began to increase as one could call to get the time and weather and to access emergency services, so the telephone network became yet more valuable. Phone machines, faxes, and cell phones once again increased the power and number of terminals, making the value soar. Eventually, everyone will have a cell phone and fax, and the growth in the terminals will slow to the population growth; however, the "power" of the phone-terminal will still continue to increase with streaming video, Internet access, and other functions limited only by the reader's imagination.

Another non-telecom example is that of highways and roads. The value (traffic) of a road rises nonlinearly with regard to the number of intersections (places that you can turn) and/or the number of homes or businesses along the road.

The second statement above is a modification of Metcalfe's law, by Gilder, to reflect the processing power of the terminals on the network. Gilder states, "As the power and storewidth of terminals rises, the value of reaching them soars. Internet traffic is a good proxy for this effect."[1]

This rule is closely related to that of Rent's rule (see Chap. 3).

References

1. G. Gilder, *Telecosm*, The Free Press, New York, p. 265, 2000.
2. J. Westland, "The Growing Importance of Intangibles: Valuation of Knowledge Assets," *BusinessWeek CFO Forum*, 2001.
3. http://searchnetworking.techtarget.com/sDefinition/0,,sid7_gci214115,00.html.

SAVINGS USING ALL-OPTICAL COMPONENTS

Cost savings to replace optical-to-electrical-to-optical (OEO) with pure optical pass-through can be estimated as

$$S_{\%} \approx X\lambda_{\#}$$

where $S_{\%}$ = percentage of savings for $\lambda_{\#} < 32$ and $\lambda_{\#} > 4$
   X = a scaling constant depending on the state of art and market forces, generally between 2 and 3
   $\lambda_{\#}$ = number of DWDM wavelengths

Discussion

It has long been known that enormous cost savings can be realized by using all optical (optical-to-optical-to-optical, OOO) switches, routers, add/drops, and transponders, as compared with conventional OEO components. The requirement to convert the optical signal to electrical, perform a function on it, and then convert back to optical is simply inefficient. It requires large amounts of power and real estate as well as large provisioning costs.

Buescher[1] studied the relative cost savings available by using optical pass-throughs on synchronous optical network (SONET) rings rather than electrical transponders. He found that the relative cost savings is quite linear from about 4 to 32 lasers. The equation to describe this line is roughly what is presented in the rule above. One could assume that for DWDM systems with more than 40 wavelengths, the cost would be between 90 and 99 percent, increasing asymptotically.

Some[2] have noted that photonic switches reduce the switch capital expenses by 60 percent as compared to OEO components. Others, such as Jagdeep Singh,[3] have noted that a single all-optical chip can reduce the cost for a optical network by a factor of two, while a factor of ten is the goal.

Although about 700 or so companies are working on OOO components, the first all-optical chip was reported in early 2002 by Infinera.[3]

References

1. B. Buescher, "Benefits of Metropolitan Mesh Optical Networks," *Proceedings of the National Fiber Optic Engineers Conference*, 2000.
2. L. Ceuppens et al., "Photonomics, the Economics of Photonic Networks," *Proceedings of the National Fiber Optic Engineers Conference*, p. 178, 2001.
3. O. Malik, "The Light Brigade," *Red Herring*, April 2002, pp. 40–44.

TELCO HOTEL COSTS

Rental rates for equipment space in a telco hotel runs approximately $100 to $200 per square foot per month.

Discussion

Active fiber links require switches, add/drops, amplifiers, and a myriad of support equipment. Generally, these devices are housed in a building with the proper air conditioning, heating, on-site service, and battery back-ups. Sometimes, the "hotel" is owned by a third party that leases space to multiple network providers or other telecom businesses. During the build-up of the late 1990s, floor space in these telecommunication equipment storage sites was at a premium, and costs soared to $200 or more per square foot.

"Just to provide some idea of recurring space and power costs, space costs can range from $500 to $1000 per seven-foot rack per month, while power costs can range from $7 to $12 per amperage per month. These costs do not include any setup or one-time costs such as rack mounting or power connection."[1]

Dense wavelength division multiplexing (DWDM) has the ability to route significantly more data though existing or new fibers, so there is a great savings in the deployment of fibers. However, the price to be paid for DWDM is in the equipment to activate the fiber links. With the possible exception of amplifiers, which may be in the field, this equipment generally resides in climate-controlled telco hotels. When considering the charm of DWDM, don't consider only the initial cost of the equipment, but also the recurring cost of the building and utilities to support it. Often, time division multiplexing (TDM) involves lower economic demands on floor space.

As Redifer relates,

New building costs for telco type applications are approximately $145.00 to $175.00/sq. ft, while costs in high-rise office buildings in major metropolitan areas may be as much as three or four times higher. WDM of multiple OC-48 systems will require four times as much floor space in the POPS as the TDM equipment. The TDM equipment will require more space at repeater sites, but the cost at these sites is generally much less than at the POPS. Typical battery plant cost for a 10,000-amp facility is approximately $1,500,000.00. Dense WDM of OC-48 systems require more amperage, breaker positions, and cabling than the equivalent capacity OC-192 system.[2]

This high cost to house, power, and maintain equipment gives economic impetus to the development of smaller and smaller devices. New generation, all-optical and integrated network elements will be much smaller and more power efficient than their predecessors, and hence they will have a much lower life-cycle cost. One example related to the authors of a customer who exhibited great interest in a nonhermetic package—just to shave half an inch off the length of a modulator.

References

1. P. Wong, "Network Architectures in the Metro DWDM Environment," *Proceedings of the National Fiber Optic Engineers Conference,* 2000.
2. G. Redifer, "DWDM vs. TDM in Metro Long-Haul Applications," *Proceedings of the National Fiber Optic Engineers Conference,* 1997.

BANDWIDTH ELASTICITY

A one-unit decline in bandwidth price yields a five-unit increase in demand.

Discussion

Price elasticity is a fundamental principle of economics. A highly elastic commodity has its demand closely related to its price, while an inelastic commodity's demand has little relationship to its price. Necessities such as food and oil tend to be inelastic; consumers will pay what is asked and only reduce demand slightly due to an increase in price. Conversely, luxury items, such as pay-per-view movies, fur coats, and caviar, tend to have a demand that changes drastically as a function of the price.

Gilder seems to indicate that bandwidth is highly elastic, and reducing the cost results in a much greater usage. It is not clear if the corollary would also hold true—that increasing the cost of bandwidth would result in decreasing demand with a slope of 5. This is because increasing bandwidth can be considered an instantaneous luxury, but once increased it becomes more of a necessity. When the increase in bandwidth becomes entrenched in a company or individual, it may be viewed as a "necessity," and the consumer will pay more for it. It is possible that bandwidth is highly elastic when prices drop and highly inelastic when prices increase—an interesting business model for sure.

Gilder states, "Fears of the bandwidth glut are groundless. Bandwidth multiplies its own demand. (As one example, business-to-business bandwidth has been exploding, and price per bit plummeting, and business-to-business traffic on the UUNet has been doubling every 90 to 120 days)."[1]

Reference

1. G. Gilder, *Telecosm*, The Free Press, New York, p. 267, 2000.

THINK STANDARDIZATION

When designing packages for electro-optic components, try to standardize using existing footprints and interfaces.

Discussion

The author of the reference wisely counsels the use of design-for-manufacturing-and-automation (DFMA) principles. He recommends the use of standards (always important) and that the issue of component handling be considered right from the start. This is particularly important with small, sensitive optical and electrostatic devices. Having to work within a small electronics package is difficult. Avoid that challenge by assembling as many of the components as possible outside of the package and minimizing inside-the-package manipulations.

In addition, it is important to properly design the handling hardware in the early stages of development. By placing appropriate fiducial marks on packages, a machine can do the precision placement of components. In addition, building a control loop around the part to be positioned allows for automatic precision location. Since the optical positioning is usually the critical factor in design success, automated location technology can be a big help. In addition, since parts always move, make sure that parts require careful positioning with respect to one another are all mounted on the same rigid structure.

Reference

1. Randy Heyler, "Packaging the Goods," *oemagazine*, 1(10), p. 32, 2002.

Electronics and MEMS

While it is possible to conceive of an approach to optical telecom systems that uses large lasers, complex and power-consumptive switches, and other nonminiaturized components, today's hybrid approach to system development is clearly on the right path. It would be prohibitively expensive to base conventional large macro-optics in telco hotels and switching stations. One can only imagine the size and complexity of a conventional (non-MEMS) optical bench to switch 300 laser lines from a cable of just 24 fibers to another cable of 24 fibers, or a spectrometer to monitor 300 laser lines from multiple fibers. By employing MEMS for doing the optical work on integrated circuits, the size is kept reasonable, and the speed is kept high. Moreover, this is on the path to the optical-optical-optical (OOO) future that everyone imagines. OOO describes a technology in which the conversion from optical fibers to electronic circuits and back to optical fiber will give way to systems in which all switching and system functions are conducted in the optical regime. We are witnessing a consistent reduction in power consumption, volume, and cost, while bandwidth, spectral purity, reliability, and overall performance continue to improve. Much of this progress is the direct result of adapting technologies that are being developed for other silicon applications to telecom systems. Chief among these are integrated circuits (both analog and digital), switching and amplifying components, semiconductor laser transmitters, detector arrays, and integrated transmitter/receiver devices. At the same time, the miniaturization of not only electronic circuits but also mechanical devices is getting substantial attention for commercial, medical, space, and military applications. The optical telecom designer benefits from all of these advances.

Miniaturization, power consumption, and reliability are so critical to dense wavelength division multiplexing (DWDM) systems that microelectromechanical systems (MEMS), epitaxial growth techniques (e.g., MOCVD), and photolithography are not only enabling technology, they also are necessary technologies for profitable deployment. Hence, billions of dollars of venture capital has flown into related activities, blurring the distinction between the optical and electrical by employing MEMs, wafer-scale IC processing (for electronic and mechanical devices), and innovative integrated circuit processing on materials other than silicon (especially, GaAs, InGaAs, and other semiconductor laser materials). Products are available with miniature mechanical and optical components to perform switching, spectral analysis, optical power monitoring, and precise fiber alignment. The integration of arrays of laser transmitters and arrays of gigahertz detector receivers for repeaters is in development. Eventually, MEMs and IC technology will allow these integrated transmitter/receiver devices to become "smart" and employ their own spectrometer for optical monitoring.

Three-dimensional integrated circuits are another intriguing technology that has been in the development since the 1970s and now, finally, promises to become available for optical telecom. Several research avenues are being explored to build chips on chips to eliminate the interconnect delays and reduce real estate. Interestingly, this technology may allow Moore's law (see Chap. 2, p. 52) to continue once the quantum limit is reached.

In this short chapter, we include a few rules and ideas that should be kept in mind during the component selection and system design process. Some of the ideas relate to the protection of devices from environmental damage, such as excessive thermal extremes, while others provide insight into the voltages required to operate micromirrors. Thermal control concerns are of importance in a number of ways in the specification, selection, and implementation of fiber systems, and the rules in this chapter are complemented by others in other chapters. One key rule that relates to estimating yields based on defects and processing steps is important to keep in mind whenever considering the use of a MEMs device, ASIC, focal plane array, semiconductor laser, or new wafer-scale photolithographic process. Many complicated yield models are based on those of Murphy and Seeds, and the reader can quickly use these to estimate and challenge the yield of the semiconductor laser manufacturer. At the same time, we have concentrated herein on concerns about the heat generated from circuit boards, its removal, and related issues.

PHOTOLITHOGRAPHY YIELD

Chip yield (be it a detector, ASIC, semiconductor laser, or MEMs device) is inversely proportional to the size of the chip and defect density raised to the number of masking steps.

Discussion

The yield of a photolithographic process is proportional to

$$\text{Yield} = \frac{1}{(1 + DA)^m}$$

where   Yield = expected fraction of successful die from a wafer
D = expected defect density per square centimeter per masking step (If unknown, assume one defect per square centimeter for silicon, more for other materials.)
A = area of the chip in square centimeters
m = number of masking steps

Moreover, Murphy developed the following model for the probability that a given die is good from a wafer:[3]

$$P_g = \left(\frac{1 - e^{-AD}}{AD}\right)^2$$

where P_g = probability that a die is good after the processing of a wafer

Additionally, Seeds give us:[2]

$$P_g = e^{-\sqrt{AD}}$$

Calculating yield is a complicated process; foundries typically have complex models to do this. Yield is also sensitive to design layout, design methodology, type of circuit, production line quality, and the age of the manufacturing equipment. However, the above equations provide a good approximation with the assumptions that the process is well controlled and developed, crystal defects are low, and breakage and human handling do not contribute. Often, this rule can be used to estimate complete device yield as steps like sawing the wafer and packaging usually are of high yield.

One must use the right numbers for defect density. This estimation isn't always trivial, as it can depend on not only the obvious (rule size, material, material purity) but subtle process attributes (individual machine, operators, the kind of night the operator had before, and even room temperature and humidity). It is probably best to scale these yields with a constant based on experience in producing a similar IC. For a more exotic material than silicon (as is often used with optical telecom lasers and detectors as well as MEMS devices), the defect density is usually higher.

Technology is always getting better, so the form of these equations may change as processes improve, as implied by the lack of masking steps in the second two equations.

Some defects can cause a circuit to be inoperable but do not affect the interconnects. Generally, yield is better when considering the fact that some of the chip is just made up of interconnects, and by ignoring the edges of the wafer (where circuits are not fabricated and crystal defects are larger).

Often, the exotic materials used by EO engineers have wafer sizes much smaller than those of silicon (e.g., the common 3-in dia. for InGaAs as opposed to 8- or 14-in dia. for silicon); therefore, the wafer size and impacts on final product yield should be considered for these materials. The cost per square centimeter for Si is not likely to decrease much in upcoming years. Future reduction in Si chip cost is more likely to occur from decreased feature size, added functionality per chip, and increased wafer size rather than reduced wafer processing costs.

References

1. R. Geiger, P. Allen, N. Strader, *VLSI Design Techniques for Analog and Digital Circuits*, McGraw-Hill, New York, pp. 19–27, 1990.
2. R. Seeds, "Yield and Cost Analysis of Bipolar LSI," *IEEE International Device Meeting*, p. 12, 1967.
3. B. Murphy, "Cost Size Optima of Monolithic Integrated Circuits," *Proceedings of the IEEE*, 52, pp. 1537–1545, Dec. 1964.
4. J. Miller, *Principles of Infrared Technology*, Van Nostrand Reinhold, New York, pp. 124–125, 1994.
5. R. Jaeger, *Introduction to Microelectronic Fabrication*, Addison-Wesley, Reading, MA, pp. 167–169, 1993.

FREQUENCY ROLL-OFF

The point at which the transimpedance amplifier gain is down 3 dB is generally

$$f_{-3dB} = \frac{\sqrt{f_{rc}f_t}}{2}$$

where f_{-3dB} = frequency at which the transimpedance amplification has fallen by 3 dB

f_{rc} = RC corner frequency = $\dfrac{1}{2\pi R_{Feedback}(C_{Diode} + C_{Input})}$

f_t = amplifier's open loop gain-bandwidth product

Discussion

Generally, a detector will be attached to a transimpedance amplifier using some feedback resistor, $R_{Feedback}$. The frequency response of this amplifier is critical in modern high-bandwidth telecom systems. It is generally given that the output starts to roll off at $F_{rc}= 1/(2\pi RC)$, where R is the resistance and C is the capacitance. For responses in the tens of gigahertz, clearly, R and C must be kept as low as possible. However, as described by Hobbs,[1] an additional $\sqrt{2}$ to 2 is lost. The −3 dB position of the roll-off can be calculated using the above equation, which is normalized to the unity-gain crossover frequency. The above does not account for noise.

Reference

1. P. Hobbs, *Building Electro-Optical Systems, Making It All Work,* John Wiley & Sons, New York, p. 627, 2000.

MAXIMUM PIEZORESISTIVE STRESS

The maximum piezo stress that can be included in a MEMS device is

$$\sigma = \frac{0.3078 PL^2}{t_2}$$

where σ = maximum stress and hence electrical output range
 P = pressure
 t = diaphragm thickness
 L = length

Discussion

MEMs optical communication devices often employ piezo devices for a myriad of applications including micromirror switches, miniature Fabry–Perot analyzers, microscanning of arrays (or dither), and fiber positioning.

Sparks[1] points out that the process used to etch out the cavity must be taken into account during the design process. He notes that the designer should be careful to properly consider the following:

- Diaphragm thickness and area
- Resistor placement with respect to cavity edge
- Resistor junction depth and doping level
- Die attachments and packaging material

Reference

1. D. Sparks, T. Bifano, and D. Malkani, "Operation and Design of MEMS/ MOEMS Devices," in P. Rai-Choudhury, Ed., *MEMS and MOEMS Technology and Applications*, SPIE, Bellingham, WA, p. 77, 2000.

MECHANICAL PROPERTIES OF TORTIONS

The following formula allows an estimation of the spring constant for a suspended MEMS mirror.

$$K = 2\beta G \frac{a^3 b}{L_c^2 L_T}$$

Yao and MacDonald[1] report that β, which is determined by the shape of the hinge, is about 0.3. The parameters a and b are the width and height of the tortion bar, and L_T is the length of the bar. L_C is half of the length of the mirror.

Discussion

The use of MEMS mirrors is desirable in many network elements such as switches, combiners, add/drop devices, and multiplexers. One technique for routing signals is to employ small tortion bars to control the position of the mirror. For typical values, K will be about 10^6 N/μ. This large number is due to the high sheer modulus of the material, Si<100>, about 5×10^{10} N/m^2.

Figure 3.1 illustrates a typical mirror of dimension 300 by 240 μ. One can expect a deflection of the edge of about 2 to 5 μ using a 40-V command and using a comb actuator as described in the reference.

Reference

1. Z. Yao and N. MacDonald, "Single Crystal Silicon Supported Thin Film Micromirrors for Optical Applications," *Optical Engineering*, 36(5), p. 1409, May 1997.

Figure 3.1 Typical suspended MEMS mirror.

MAGNETOSTRICTIVE EXPANSION AND CONTRACTION

Rare-earth alloys have a maximum magnetostrictive (MS) elongation of 1000 to 2000 parts per million (ppm).

Discussion

Elemental iron exhibits a comparatively small contraction of –8 ppm. When developing a micromachine, the amount of contraction and elongation possible through an MS effect is often important for films and microstructures. Many ferromagnetic materials exhibit a contraction or elongation when exposed to a magnetic field. The above rule gives some approximations that can be expected when employing such. There are hundreds of sources on this topic on the Internet and in journals and books.

Magnetostrictive materials are finding wide use as sensors and actuators in a diverse range of applications. The material properties cause the size of the component to change when exposed to a magnetic field. They are not as efficient as piezoelectric components, but this is offset somewhat by their durability. At the same time, when an MS material is exposed to mechanical strain, its magnetic properties change. Because of their ability to both sense and actuate, these materials are used in smart (self-correcting) structures such as ultrasonic vibrators.

The two MS effects are named for the scientists who discovered them. The Joule effect describes the change in size of a material when exposed to a magnetic field. The opposite effect, in which a stress applied to the material causes a change in its magnetization, is called the Villari effect (also referred to as the magnetomechanical effect).

These materials also can be distorted in unexpected ways. By passing a current through the material, a spiral magnetic field and associated torsion of the material can be created. This is called the Wiedemann effect.

Practical application of MS effects began over 100 years ago, using materials with a modest response in size change per unit length. By the 1970s, "giant" effects had been found that create changes as large as 1000 ppm.

Reference

1. D. Sparks, T. Bifano, and D. Malkani, "Operation and Design of MEMS/MOEMS Devices," in P. Rai-Choudhury, Ed., *MEMS and MOEMS Technology And Applications*, SPIE, Bellingham, WA, pp. 53–54, 2000.

MICRO-OPTICS CROSSTALK

The crosstalk (χ) of beams in a micro-optics MEMS optical fiber switch can be expressed as

$$\chi = 10\log\left[e^{-(\theta^2/\theta_o^2)}\right] = -4.43\left(\frac{\theta}{\theta_o}\right)^2 \text{ dB}$$

where θ_o = numerical aperture of the switched beam $(\lambda/\pi\omega_o)$ and ω_o = its beam waist

θ = angle through which the beam is deflected

Discussion

Some modern switching systems for fiber networks employ angle deflection of the beam emitted by one fiber and its subsequent reception by another of a number of fibers. This scheme is shown in the Figure 3.2.

This result is derived from the fact that the deflected beam is a Gaussian cross section. It also assumes that the possible output fibers are abutting so that any pointing error by the deflection switch will put some energy into the wrong fiber, which is another way of saying that there will be some crosstalk.

Using the above equation and a goal of achieving –50 dB isolation between adjacent channels, we find that the deflection angle should be 3.4 times as large as θ_o.

As an example, for a wavelength of 1.355 µ and a fiber with a core diameter of 0.1 mm, the numerical aperture is 4.3 milliradians.

Reference

1. S. Glockner et al., "Piezoelectrically Driven Micro-Optic Fiber Switches," *Optical Engineering*, 37(4), p. 1229, April 1998.

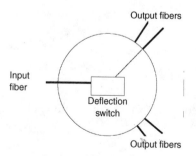

Figure 3.2 Micro-optics insertion loss.

DEAL'S OXIDE GROWTH MODEL

For short oxidation times, the growth of the thickness of an oxide layer (d) is linear and follows the following equation:

$$d = \frac{B}{A}(t + \tau)$$

where t = deposition time
τ = time it takes to accommodate the initial oxide thickness

For long oxidation times, the following parabolic law can be assumed:

$$d = \sqrt{Bt}$$

Discussion

The distinction between short and long times is made as follows:
- Short time = $t \ll A^2/4B$
- Long time = $t \gg A^2/4B$

A, B, and other parameters are defined below.

$$A = 2D\left(\frac{1}{k} + \frac{1}{h}\right)$$

$$B = \frac{2DC^*}{N_1}$$

$$\tau = \frac{x_i^2 + Ax_i}{B}$$

The additional definitions are as follows:

x = SiO$_2$ thickness
k = reaction rate constant
h = gas phase transport coefficient
C^* = solubility of oxidant in SiO$_2$
D = diffusion coefficient of oxidant in SiO$_2$
x_i = initial thickness of oxide
N_1 = number of oxidant molecules needed to form 1 cm^3 of SiO$_2$ (10^{22} for O$_2$ and 2×10^{22} for H$_2$O)

The following table summarizes some of the key parameters as a function of temperature for dry oxidation of Si.

Temperature (°C)	B/A (microns per hour)	τ (hours)
700	0.00026	81
800	0.003	9.0
920	0.028	1.4
1000	0.071	0.37
1100	0.3	0.076
1200	1.12	0.027

References

1. M. Madou, *Fundamentals of Microfabrication*, CRC Press, Boca Raton, FL, 1997.
2. C. Gorecki, "Optical Waveguides and Silicon-Based Micromachined Architectures," in P. Rai-Choudhury, Ed., *MEMS and MOEMS Technology And Applications*, SPIE, Bellingham, WA, pp. 219–221, 2000.
3. B. Deal and A. Grove, "General Relationship for the Thermal Oxidation of Silicon," *Journal of Applied Physics*, 36, 3770, 1965.
4. Lecture on oxidation of silicon in the curriculum on "Semiconductor Material Processing" at Rensselaer Polytechnic Institute, http://banyan.cie.rpi.edu/~class/SMP/oxidation.pdf.

ARRHENIUS EQUATION APPLIED TO ELECTRONICS

It is quite common for accelerated testing to make use of the Arrhenius equation, which takes advantage of the effects of temperature and current. A general form of the equation is

$$\text{MTTF} = AJ^{-2}e^{E_q/kT}$$

where   MTTF = mean time to failure

        A = a constant that is specific to the item being tested (found by calibration using a large sampling of items)

        E_g = characterization of the energy of a process (for example, electromigration of the components of a polymer)

        J = current density in A/m^2

        k = Boltzmann's constant

        T = temperature in kelvins

Discussion

Typical operational environments produce a gradual degradation that is too small to be used in predicting operational life. By adjusting T or J, the system can be exposed to an equivalent lifetime much larger than occurs at normal operating temperature.

Examples of use of the Arrhenius equation abound on the World Wide Web. The reader might want to examine Ref. 2 for details.

To use the above equation for estimating the accelerated aging of components, you must use Boltzmann's constant in units of eV/K. The constant is about 1.23×10^{-23} J/K or 8.6×10^{-5} eV/K.

As an example of reliability, some switches used in optical networks take advantage of the thermo-optic effect in polymers. Of course, as in all component networks, reliability is important. The authors of the referenced paper[1] performed accelerated testing on polymer switches. Using spectroscopic methods, they monitored the changes in the chemical composition of the polymer during the elevated temperature experiments. Even at temperatures around 125°C, the rate of change of the material is so slow that component lifetimes of 100 years could be expected. The Arrhenius-type model provides a good model of the reliability that can be expected once the proper constants are determined.

It should be pointed out that not everyone agrees that all failure mechanisms can accurately represented by this type of model. This type of model should be used only after it has been confirmed that an activation energy model makes sense.

References

1. P. De Dobbelaere et al., "Intrinsic Reliability Properties of Polymer Based Solid-State Optical Switches," *Proceedings of the National Fiber Optic Engineers Conference,* 1998.
2. "What Causes Semiconductor Devices to Fail?" by the staff of *Test & Measurement World,* Nov. 1999, http://www.tmworld.com/articles/11_1999_models.htm.
3. R. Jaeger, *Introduction to Microelectronic Fabrication,* Addison-Wesley, Reading, MA, pp. 33–36, 1993.

STONEY'S FORMULA FOR STRESS FROM A DEPOSITED FILM

The stress of a deposited film and resulting curvature are given by

$$\sigma = \pm \left| \frac{E}{6(1-v)} \frac{d^2}{t} \left(\frac{1}{R_A} - \frac{1}{R_B} \right) \right|$$

where E = Young's modulus of the substrate
v = Poisson ratio of the substrate
R_A = curvature radius after deposition
R_B = curvature radius before deposition
d = film thickness
t = substrate thickness

Discussion

In micromachining, filter manufacture, detector manufacture, and semiconductor processing, a thin film is typically deposited on a substrate. Because they are different materials with different thermo-mechanical properties, tensile and compressive forces result. The rule is based on simple physics exploiting the ratio between Young's modulus and the Poisson modulus.

For most films, one can assume that the deposited film's thickness is much less than the substrate's thickness, and such is implicit in the above equation. In fact, Gorecki[2] indicates that d/t is 200 to 1000 for thin films on silicon. The term $E/(1-v)$ is simply the biaxial modulus of the substrate.

Also as pointed out in Ref. 2, stress in silicon dioxide depends on dopant concentration, porosity, water content, and deposition parameters. Also, stress in silicon nitride is mostly compressive and decreases quickly with increasing oxygen content.

References

1. M. Madou, *Fundamentals of Microfabrication*, CRC Press, Boca Raton, FL, 1997.
2. C. Gorecki, "Optical Waveguides and Silicon-Based Micromachined Architectures," in P. Rai-Choudhury, Ed., *MEMS and MOEMS Technology And Applications*, SPIE, Bellingham, WA, pp. 232–233, 2000.
3. A similar result is found at http://www.frontiersemi.com/appintro.shtml.

NTSC VIDEO

When using fiber for transmission of video, use a modulation index (μ) of 0.25 to 0.33.

Discussion

For conciseness and completeness, we cannot improve on the following quote from Effenberger:[1]

> Standard practice has been to set the rms modulation index to be 0.25 to 0.33, which is equivalent to setting the clipping threshold 4 to 3 standard deviations away from the mean of the signal distribution, respectively. For a numerical example, a system that delivers 50 channels of AM video with a μ of 0.25 would require a received power of –9 dBm. (Assuming noise power = –106 dBm, optical modulation = 0.05, and receiver responsivity = 7.36 $W^{-1/2}$.

Reference

1. F. Effenberger, "Efficient Optical Transmission of NTSC Video," *Proceedings of the National Fiber Optic Engineers Conference,* 1996.

Fiber Mechanical Considerations

Optical fibers, ideally, would be immune to environmental influences, mechanical stresses, and other effects. As is always the case, the system designer cannot ignore these issues. This chapter is included to provide some insight into the practical problems that can develop with these very long, thin slivers of glass. Finally, fibers have been deployed in telecom cables for decades, and statistics and lessons learned about the real-world mechanical properties are brought to light (pun intended) herein.

The rules in this chapter provide some information on how fibers react to mechanical stress from bending or cleaving, exposure to humidity, and other effects. For example, stress or temperature change in a fiber can change the wavelength properties of gratings built into the fiber. The interested reader will note that this topic is also discussed in other chapters of the book. Another factor in the performance of fiber optics includes the mechanical history of the installed system. One of our rules advises that a requirement for testing should be to ensure that "in-field" use conditions will be within the expected guidelines. We include in this topic the maximum stress that the fiber should experience before, during, and after installation. Interestingly, the strength of the fiber is affected by its stress history.

Most fiber mechanical failures derive from crack propagation and interaction of the fiber with the strength members in the cable. We address these issues, concentrating on quantitative expressions for the relationship between stress and a reduction in component life. At the same time, mechanical failures are a desirable feature of the fiber, since they allow the fiber to be cleaved in a more or less predictable manner.

A similar issue is encountered when the designer builds bends into the fiber for interface with components and cables. Limitations to the bend radius are critical and are addressed. A similar rule addresses the performance of connectors that are exposed to contact deformation and how they react when the connection is broken.

Humidity effects in fibers draws attention in this chapter. Strength is adversely affected, and humidity can alter the properties of the fiber itself, as well as those of other optical elements. In addition, moisture can lead to corrosion problems in metallic components of the network. Of course, it is widely known that water in the fiber (even in trace amounts) leads to important attenuation features in particular bands. Test methods are included in several of the rules. At the same time, one always wants to be able to predict performance, especially in complex systems. One rule deals with this very issue, offering a statistical model.

Attention is given to temperature cycling of the fibers and some failure mechanisms that can emerge for systems exposed to varying temperatures. Finally, we take note of a point made by Girard, who expresses concern about the performance of fibers that carry large power loads. This can occur when a Raman pump or amplifier is used, resulting in as much as 2 W being propagated in a single-mode fiber.

BENDING A FIBER

Optical fibers are flexible, with curvature radii 20 to 50 times their diameters when hot (during manufacture). After cooling, they are bendable to within about 300 diameters.

Additionally, Ronan[1] suggests that a glass optical fiber should not be coiled on a diameter less than 600 times the fiber diameter for long-term use, or tighter than 200 times momentarily.

Discussion

In practice, it is rare to see spools smaller than about 10 cm dia. Moreover, at longer wavelengths (in the vicinity of 1625 nm), sensitivity of transmission to bending can be quite severe. This rule is important, since network monitoring is often done at this wavelength. At least one manufacturer indicates that, if its single-mode fiber is bent into a single 32-mm loop, up to a 0.5 dB additional attenuation can occur. (This is a ratio of loop to fiber core diameter of about 3000). In contrast, 100 loops at 50 mm produce no more than 0.1 dB of attenuation. The message is clear: *avoid small loops!*

Bending optical fibers can reduce their life. During bending, the center of the fiber is stressed. Bending tighter than these curvatures will damage the fiber. The inner edge of the fiber is under compression, while the outer edge is under tension. The tension causes microcracks on the surface of the glass to grow, and this weakens the fiber.

Ignore this rule at your own peril. Otherwise, you might end up with two pieces of fiber when you previously had only one.

This rule is useful anytime your system requires bends in a fiber. Bending more than specified above will often result in overstressing the glass, which may result in a future break. It is generally accepted that when a fiber is bent, you must ensure that the radius of bend curvature is more than 500 (a safety margin) times the fiber diameter.

The attenuation of light in a fiber that has tight bends is greater for higher wavelengths (e.g., 1.55-μm light has more loss than 0.85-μm light for the same bend radius).

Reference

1. Private communications with G. Ronan, 1995.

CLEAVING FIBERS

To produce a nearly flat cleave of an optical fiber, the crack speed must propagate slower than the speed of sound in the fiber.

Discussion

A cleave is initiated by lightly scratching the surface of the fiber. When the fiber is thereafter pulled or bent, a crack will originate at the scratch and propagate radially across the width of the fiber. The stress field within the fiber created by tension or bending determines the speed at which sound will propagate. If the crack exceeds this speed, the crack will suddenly change direction by almost 90°. This results in an excess of glass on one fiber (commonly called a *hackle*) and a shortage on the other fiber.

This rule should hold regardless of fiber material. Just be certain that you know the speed of sound in the fiber. Keep in mind that most bad cleaves are due to the initial scratch being too deep. Torsion will not change the speed of sound within the fiber, but it will produce nonperpendicular endfaces.

It is easy to cleave an 80-µm dia. fiber, possible to cleave a 125-µm dia. fiber, and usually difficult to cleave >200-µm fibers. To some extent, the difficulty in cleaving these fibers results from the fact that the material of the fiber is not crystalline. Again, torsion will produce a nonperpendicular endface. In fact, most commercially available angle cleavers rely on torsion. The endface angle is proportional to the amount of torsion.

References

1. Private communications with G. Ronan, 1995.
2. Product catalog, Wave Optics Inc., Mountain View, CA, p. 23, 1995.

DIFFUSION OF WATER THROUGH LIGHTGUIDE COATINGS

The loss of strength of fibers exposed to humidity can be estimated by the following equation:

$$\sigma = A + B \exp(-RH \times C)$$

where σ = strength in GPa
RH = relative humidity as shown in Figure 4.1.

Typical values are $A = 6.002$, $B = 2.1795$, and $C = 2.874$.

Discussion

Humidity is the enemy of fibers. Its effect is to reduce the strength of the fiber by reducing the energy required to initiate crack propagation. Another problem is pitting due to dissolution of silica on the surface of the fiber. In any case, it is well known that the strength of the fiber is highly correlated to the humidity within the fiber coating. Not only are fibers exposed to stress adversely affected by the penetration of water vapor, but even fibers that are exposed to no stress sustain accelerated aging.

Even overnight exposure to elevated humidity levels can be a factor. In fact, the diffusion time is short (about 10 min), so even temporary exposure can be an issue. Fortunately, for most coatings, the exit of the water

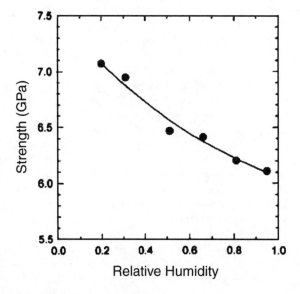

Figure 4.1   Fiber strength vs. relative humidity.

from the fiber/coating system occurs on a scale equal to the uptake time, once the external environment is dry. The choice of coating material is critical as well. Success is achieved with coatings that are impermeable, but the most common coatings allow water to flow into and out of the fiber/ coating environment.

In addition to loss of strength, it is important that the conditions during manufacture be such that water is not introduced into the fiber or the prominent water absorption feature at 1385 nm. Because of the importance of this topic, research continues, and new results are likely to guide the selection of both fiber and coating materials.

References

1. C. Kurkjian, J. Armstrong, M. Matthewson, and I. Plitz, "Diffusion of Water through Lightguide Coatings," *Proceedings of the National Fiber Optic Engineers Conference*, 1996.

2. J. Mrotek et al., "Diffusion of Moisture through Optical Fiber Coatings," *IEEE Journal of Lightwave Technology*, 19(7), July 2001.

FIBER LIFETIME STRESS INTENSITY FACTOR

Set the test extremes at 2× the normal operating conditions.

Discussion

The basic process for the manufacture of a fiber and the construction of a cable should be carefully examined to ensure that it isn't producing undue stress and damage on a fiber. As pointed out by Glaesemann,

> In most cases, fiber-handling equipment is qualified after engineers have carefully tuned the equipment so that it is examined under the best of conditions. However, to test processing equipment for possible failure modes, one would run the equipment at its extremes for speed, alignment, and stress.[1]

We also note that the Arrhenius equation (mentioned in Chap. 3) defines the accelerated aging of all classes of systems as a function of temperature.

This rule should not surprise anyone who has worked in situations where failure is expensive. In fact, in most engineering applications, this rule is modified to change the "2" to a somewhat higher number—say 5 to 10. For example, it is not uncommon for elevators to use four to six cables, each of which is able to carry the car by itself.

Reference

1. G. Glaesemann, "Advancements in Mechanical Strength and Reliability of Optical Fibers," *Proceedings of the SPIE Conference on Reliability of Optical Fibers and Optical Fiber Systems,* Sept. 20–21, 1999. Also available at www.corningfiber.com.

FIBER STRENGTH TESTING

According to Camilo and Overton,[1] "The strength of modern optical fibers is defined by very small defects localized in the surface of the glass."

Discussion

Advances in testing the strength of fibers have been significant. In the past, short sections of fiber were tested. A number of causes can be attributed to these defects, including production and coating processes. Two-point bending is a convenient method of testing the static and dynamic strength of the resulting fiber.

The strength of a fiber tested in a two-point bending system is equal to

$$S_f = E(\varepsilon_f)\, \varepsilon_f$$

where $E(\varepsilon_f) = E_0\, \varepsilon_f (1 + 2.125\, \varepsilon_f)$

and

$$\varepsilon_f = 1.198 \frac{d_i}{d_f - d_c + 2 d_g}$$

where ε_f = maximum strain on the fiber when it fails
d_i = glass fiber diameter
d_c = overall fiber diameter including any coatings
d_g = total depth of both grooves in the parallel plates between which the fiber is bent
d_f = gap between these faceplates at failure
E_0 = modulus at zero strain (72.2 G Pa)

Reference

1. G. Camilo and R. Overton, "Evolution of Fiber Strength after Draw," *Proceedings of the National Fiber Optic Engineers Conference*, 2001.

FIBER STRENGTH

Fiber strength depends on humidity.

Discussion

Figure 4.2 illustrates the loss of strength that occurs after 24 hr of exposure to a given level of humidity (50 percent) and temperature (23°C). Two-point testing was used to measure the strength. Overton et al.[1] note that "Glass strength has been shown in the past to be at 10 GPa or more in an absolutely moisture-free condition," and they observe that the mechanism for strength loss may be related to moisture in a crack that accelerates oxidation and subsequent stress. They also note that all fibers in the tested samples exhibited this behavior. The reduced surface energy in the presence of humidity reduces the stress at which crack growth can occur.

Reference

1. R. Overton, G. Camilo, M. Purvis, R. Greer, "Interface Integrity—How Material Interfaces Can Affect the Performance of Dual Coated Optical Fiber," *Proceedings of the National Fiber Optic Engineers Conference,* 1999.

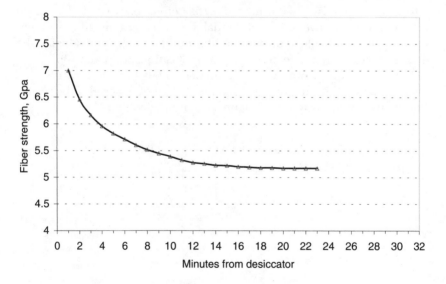

Figure 4.2   Fiber strength vs. time.

MOISTURE DAMAGE TO MICRO-OPTICS

Corrosion can be a factor in the reliability of fiber systems. For the corrosion effect, we define a corrosion factor (α) that is described by a form of the widely known Arrhenius equation.

$$\alpha = \alpha_0 RH e^{-\Delta H/kT}$$

where ΔH = activation energy in the same units as kT
RH = relative humidity factor
α_0 = material related corrosion parameter
k = Boltzmann's constant (1.38×10^{-23} J/K)
T = temperature of the device

Discussion

Elsewhere in this chapter, we have described the adverse effects of moisture in cables. Here, we include a rule that captures the impact of moisture in epoxy materials. The major impact is the deformation of the epoxy in the presence of moisture, leading to loss of adhesion. The equation above addresses another issue—the more familiar effect of corrosion of metal parts. On the positive side, substantial improvement in adhesives has occurred in the dental industry (among others) so that, gradually, the adverse impact of moisture in fiber optic applications will diminish.

These values vary from material to material, but the basic form of the equation can be expected to hold for a wide range of materials. The reader will note the similarity to the Arrhenius equation, detailed Chap. 3.

Reference

1. S. Cao, "Reliability Evaluation of Micro-Optic Based Passive Fiber Optic Components," *Proceedings of the National Fiber Optic Engineers Conference,* 1997.

STRENGTH OF FIBERS

For pure silica fibers, the modulus of elasticity E is given by

$$E = 70.31 + 3(1 + 3\varepsilon) \text{ in GPa}$$

where ε = strain

Discussion

In the setup shown in Fig. 4.3, the maximum strain on a fiber is given by

$$\varepsilon = \frac{1.1978(2r)}{D - d}$$

where r = the glass fiber radius
D = the inner diameter of the reamed hole
d = the coating diameter

The maximum stress on the fiber in two-point bending is given by

$$\sigma = \varepsilon E$$

For a fiber that is held under each nominal stress level, the relation between the stress applied the fiber and the median time to fracture of the fiber is as follows:

$$\ln (t_f) = -n_s \times \ln (\sigma_f) + \text{intercept}$$

where σ_f = each nominal stress level
t_f = the median time when the fiber fractures
n_s = the static stress corrosion parameter

This function is shown in Fig. 4.4.

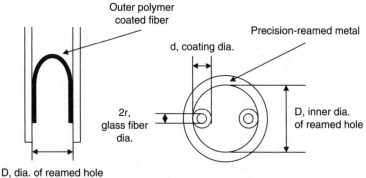

Figure 4.3 Schematic of two-point bending.

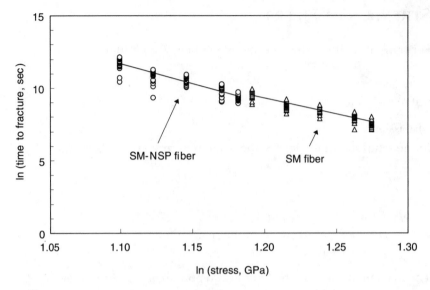

Figure 4.4   Static fatigue of unaged single mode (SM)- non-strippable primary (NSP) fiber and SM fiber

Reference

1. T. Murase, K. Shiraishi, and H. Noro, "Single Mode Non-strippable Primary Coated Fiber: The Enhancement of Mechanical Reliability," *Proceedings of the National Fiber Optic Engineers Conference,* 2000.

PACKAGING STRESS AND FIBER LENGTH

When clamping short lengths, a fiber's critical strain is acceptable if the length is equal to or smaller than the values computed below.

1. For angled clamping of fibers (relative to the exit tube),

$$l = \frac{r_0}{[\varepsilon]}\left(\beta + 60\frac{\Delta\alpha\Delta t}{\beta}\right)$$

where l = length of the fiber in a package
   ε = strain, usually 0.005 for typical telecom fibers
   $\Delta\alpha$ = difference in the thermal coefficient of expansion between the package and the fiber (e.g., 5.3×10^{-6} for a Kovar package)
   Δt = difference between the maximum short-term assembly temperature and the maximum long-term operating environment temperature
   r_0 = radius of the fiber
   β = angle that the fiber is bent to achieve

2. The maximum allowable length of a straight fiber sealed in a package, keeping the thermal strain below the critical strain, for a short exposure is

$$L_c = \pi r_0\sqrt{\frac{2(1 - 6(\Delta\alpha\Delta t))}{(\Delta\alpha\Delta t)(2 - 6(\Delta\alpha\Delta t))}}$$

where L_c = critical length, or the maximum allowable length, of the fiber inside the enclosure; at this length, the fiber buckles

Discussion

When inserting and rigidly attaching a fiber end inside a package (such as in the case of common pigtail packaging), the strain of the fiber must be considered, as it will change as a function of temperature. The strain induced to the fiber by short-term temperature increases (by soldering) or long-term thermal gradients within a pigtailed connection can be significant. This strain can result in the fiber failing at its connections or bends.

Note that, when soldering a fiber pigtail, the solder's solidus temperature is the important temperature, as this is the point at which the fiber is fixed to the package and the stress is set.

Suhir[1] goes though elaborate calculations and has useful numerical examples. These derivations assume certain boundary conditions, which are explained in his article. Be careful of the initial boundary conditions, as these can influence the actual reliability and may differ from yours.

Suhir also gives the following handy equation for estimating the stress that can be applied to a fiber as a result of short-term manufacturing procedures:

$$\sigma_t = \sigma_T \left(\frac{t}{T}\right)^{-t/n}$$

where σ_t = stress under the higher short-term manufacturing conditions
σ_T = stress under nominal long-term conditions
t = time of the short-term higher stress
T = time the fiber is expected to be used without failure at nominal stress and temperature conditions
n = a constant, usually between 18 and 20[1]

If a low-expansion thermal enclosure is used, which keeps the thermal strain low, then the critical length (l^*) in the second equation of this rule can be simplified to[1]

$$l^* \cong \frac{\pi r_0}{\sqrt{\varepsilon_1}}$$

References

1. E. Suhir, C. Paola, and W. MacDonald, "Input/Output Fiber Configuration in a Laser Package Design," *Journal of Lightwave Technology,* 11(12), p. 2087–2092, December 1993.
2. Private communications with D. McCal and G. Fanning, 2002.
3. P. Haugsjaa, "Packaging of Optoelectronic Components," Optical Fiber Communication (OFC) Symposium and Exposition short course, 2001.

TEMPERATURE CYCLING

Temperature cycling is a threat for some fibers.

Discussion

Transmission of fibers is, above all, a key feature of all systems. Therefore, we must be concerned about any threat to the integrity of the transmission. This effect usually results from thermally induced movement of the strength member of the cable, which can cause "physical changes in the internal structure of the cable which may lead to high optical loss."[1] Practice in the field suggests that proper clamping of strength members can dramatically reduce this problem.

Fulmer states,

> Wide fluctuations in temperature, particularly those involving low temperature extremes, can lead to a variety of problems in cable with no clamping of its strength member. Thermal cycling induced movement of the strength member can cause physical changes in the internal structure of the cable, which may lead to high optical loss. Even if this strength member protrusion does not result in high loss due to microbending of the fibers or some related loss mechanism (as seen in loose buffer tube cable), damage and high loss are still possible through contact of the protruded strength member with fibers/splices in the closure. Additionally, even if a strength member clamping fixture is installed on the cable, poor clamp design or incorrect installation practice can lead to bowing and, in the case of fiberglass members, eventual breakage of the strength member. Either bowing or pistoning of the strength member may lead to high optical loss despite the fact that a strength member clamping fixture is in place.[1]

Reference

1. J. Fulmer, "The Use of Proper Strength Member Clamping to Prevent High Light Loss in Thermally Cycled Fiber Optic Cable," *Proceedings of the National Fiber Optic Engineers Conference*, 1996.

WAVELENGTH SHIFT DUE TO FIBER STRAIN AND STRESS

$$\Delta\lambda = \lambda_o \varepsilon \left\{ 1 + \frac{n_o^2}{2} [P_{12} - v(P_{11} + P_{12})] \right\}$$

where $\Delta\lambda$ = wavelength shift
 λ_o = original wavelength
 ε = fiber strain in a unitless ratio of size change to initial size
 n_o = the index of refraction (assume 1.45 if unknown)
 v = Poisson's ratio (assume 0.19 if unknown)
 P_{11} = a strain optic coefficient (assume 0.113 if unknown)
 P_{12} = another strain optics coefficient (assume 0.252 if unknown)

Discussion

A force applied axially to a fiber grating will result in a wavelength shift. The P_{ij} terms are components of the photoelastic tensor. Using the suggested constants gives a relationship of

$$\Delta\lambda = \lambda_o \varepsilon \{ 1 - 0.19 \}$$

where the constant 0.19 is the effective photoelastic constant. A slightly different result comes from Ref. 2, with the constant 0.19 replaced by 0.22.

For silica, the shift is expressed as

$$\Delta\lambda = \lambda_o \varepsilon 0.78 \times 10^{-6}$$

with ε expressed in microstrain.

Connectors (especially crimp styles) also impose compressive and tensile stresses on fibers, which can cause a shift in wavelength. Although small, this shift can be on the order of the laser separation in a DWDM system with 10 or 20 GHz spacing.

This rule can also be used to estimate a Bragg grating wavelength shift. This equation can be used backward; if you measure a shift in wavelength, you can calculate the strain and, using Hooke's law ($\sigma = E\varepsilon$), the stress (σ).

Reference

1. L. Baker et al., "Fiber Bragg Gratings for Stress Field Characterization inside a Connector," *Proceedings of the SPIE Conference on Reliability of Optical Fibers and Optical Fiber Systems*, Sept. 20–21, 1999, Also available at www.corningfiber.com.
2. Roberta C. Chaves et al., "Strain and Temperature Measurements in Power Systems with Multiplexed Fiber Optics Bragg Grating Sensors," *Journal of Microwaves and Optoelectronics*, 2(1), p. 54, June 2000.

FIBER STRESS DERATING

The maximum applied stress to a fiber should be no more than 1/5 of the proof stress. If you don't know what the proof stress is, keep stress levels below 20 kpsi.

Discussion

To minimize stress-related failures, the stress applied to a fiber should be less than the proof stress (σ_p). Pulling or bending a fiber induces stress. Allowable stresses during cable manufacture and installation may be higher, as they are of short duration. Following this rule essentially guarantees zero stress-related failures for 40 years. Most fibers are stress tested to 100 kpsi, so, if you don't know the stress rating, assume it to be 100, and derate accordingly. The above is for long fibers (e.g., >1 km); shorter fibers can have slightly higher stresses.

Reference

1. R. Castilone, G. Glaesemann, T. Hanson, "Extrinsic Strength Measurements and Associated Mechanical Reliability Modelling of Optical Fiber," *Proceedings of the National Fiber Optic Engineers Conference*, Denver, CO, 2000.
2. http://www.corningfiber.com.

5

Free-Space Optical Communications

While the fiber is widely used in telecom, the use of free-space optical communication is another promising communications architecture. Two applications are emerging. First, terrestrial communication for enterprise, "last mile," and "last meter" applications or for temporary applications. The second area that is getting attention is space. Space-to-space and space-to-ground applications are being explored to increase bandwidth, provide security, and reduce the size and mass of components in orbit. Interestingly, the space applications of optical communications derive much of their technology from components and methods developed for ground/fiber systems. Fortunately for the space designer, the advances in lasers, amplifiers, and multiplexers can, in most cases, be adapted to space by providing a qualification path that yields parts of appropriately high reliability and radiation tolerance.

The terrestrial free-space communication designer has a unique set of challenges. A number of factors that control system performance are not within the designer's control, including the weather (and the associated humidity, turbulence, and smog along the communication path), laser power limitations imposed by eye safety issues, and so on. The best defense in these applications is a conservative design philosophy in which path length is limited to ranges where high-quality communication is nearly always possible and a networking strategy that allows easy expansion of the system along line-of-sight pathways. In addition, there will be areas of the world where reliable application of free-space communications cannot be expected due to frequent occurrences of fog and rain.

This chapter presents about a dozen ideas that provide shortcuts and insights for the system designer. As one might expect, the physical properties of the atmosphere are an area of attention, due to their impact on beam spread, signal strength variation, beam steering, and other effects. What is included here is but a skimming of the surface of issues related to atmospheric optics. Many journals regularly present articles on this complex topic and should be consulted for a deeper understanding.

The motivation for free-space communications is strong. Adding transmit/receive stations requires no installations of cables or other equipment, as long as a line-of-sight path can be found. For urban applications, particularly for enterprise and "last-mile" solutions, within a range of a few hundred meters, free-space communications might be superior to fiber or copper. In addition, some concepts employ a narrow beam for a low probability of intercept. This provides secure communications, since no intervening receiver can pick up the signal, which is confined to a very narrow angle. This may see uses for financial and defense customers. Free-space systems can make use of existing high-speed data services and can be used when redistribution of information is a key goal of a system. These include the downtown area of a major city, a college campus, or communications between buildings in a hospital complex where images and data need to flow at high speed and where intranet services are the focus. Free-space systems are also being considered for the "last meter" applications such as for Internet connections on airplanes or in office cubical farms.

Under ideal atmospheric conditions, free-space systems routinely provide service at nearly the same rates as fibers and also allow the dense wavelength division multiplexing (DWDM) that has already been employed in fiber networks. The system designer should benefit from the basic ideas included here.

C_n^2 ESTIMATES

The index of refraction structure constant C_n^2 is a measure of the index of refraction variation induced by small-scale variations in the temperature of the atmosphere. No other parameter is so common as C_n^2 in defining the impact of the atmosphere on the propagation of light. There are several quick ways to estimate C_n^2. The easiest is that it varies with altitude in the following way:[1]

$$C_n^2(h) = \frac{1.5 \times 10^{-13}}{h \text{ (in meters)}} \text{ for } h < 20 \text{ km}$$

and

$$C_n^2(h) = 0 \text{ for altitudes above 20 km}$$

Note that $C_n^2(h)$ has the rather odd units of $\text{m}^{-2/3}$.

Discussion

$C_n^2(h)$, in which we have explicitly shown that the function depends on altitude, is critical in determining a number of key propagation issues (such as laser beam expansion, visibility, and adaptive optics system performance), and defines the performance impact that the atmosphere has on ground-based astronomical telescopes.

Each of the estimates shown below is just that—an estimate. However, for the modeling of systems or the sizing of the optical components to be used in a communications or illumination instruments, these approximations are quite adequate. Any attempt to provide really accurate approximations to the real behavior of the atmosphere is beyond the scope of this type of book, but the subject is covered frequently in the astronomical literature, particularly in conferences and papers that deal with the design of ground-based telescopes.

Beam propagation estimates rely on knowledge of C_n^2. Many of those estimation methods appear in this chapter. Although there are a number of estimates of C_n^2 that are widely used, most will provide adequate results for system engineers trying to determine the impact of turbulence on the intensity of a beam as well as other features of beam propagation.

The most widely used analytic expression for $C_n^2(h)$ is the so-called *Hufnagel-Valley (HV) 5/7 model.* It is so-called because the profile of C_n^2 results in a Fried parameter of 5 cm and an isoplanatic angle of 7 µrad for a wavelength of 0.5 microns. Beland[2] expresses the Hufnagel-Valley (HV) 5/7 model as

$$8.2 \times 10^{-26}\left(\frac{h}{1000}\right)^{10} W^2 \exp\left(-\frac{h}{1000}\right) + 2.7 \times 10^{-16} e^{-h/1500} + 1.7 \times 10^{-14} e^{-h/100}$$

where h = height in meters

W = wind correlating factor, which is selected as 21 for the HV 5/7 model

Note that the second reference has an error in the multiplier in the last term. That error has been corrected in the material presented above.

In many cases, C_n^2 value can be crudely approximated as simply 1×10^{-14} at night and 2×10^{-14} during the day.

References

1. J. Accetta, "Infrared Search and Track Systems," in *Passive Electro-Optical Systems,* Vol. 5, S. Campana, Ed., *The Infrared and Electro-Optical Systems Handbook,* J. Accetta and D. Shumaker, Ex. Eds., ERIM, Ann Arbor, MI, and SPIE, Bellingham, WA, p. 287, 1993.

2. R. Beland, "Propagation through Atmospheric Optical Turbulence," in *Atmospheric Propagation of Radiation,* Vol. 2, F. Smith, Ed., *The Infrared and Electro-Optical Systems Handbook,* J. Accetta and D. Shumaker, Ex. Eds., ERIM, Ann Arbor, MI, and SPIE, Bellingham, WA, p. 221, 1993.

ATMOSPHERIC ATTENUATION OR BEER'S LAW

The attenuation of a beam of light traversing an attenuating medium can often be estimated by the simple form:

$$\text{Transmission} = e^{-\alpha z}$$

where a = attenuation coefficient
        z = path length

Discussion

This common form is called Beer's law and is useful in describing the attenuation of a beam in the atmosphere, water environments, and optical materials. Since both absorption and scattering will remove energy from the beam, α is usually expressed as

$$\alpha = a + \gamma$$

where a = absorption per unit length
        γ = scattering per the same unit length

The equations are direct derivations of Beer's law. This law is commonly applicable to transmission problems associated with most media. The idea is that scattering and absorption remove light from a collimated beam at a rate proportional to the thickness of the path and the amount of scattering or absorbing material that is present.

The rule is derived from the fact that the fractional amount of radiation removed from the beam is independent of the beam intensity. This fact leads to a differential equation of the form $(dz)/z$ = constant. The solution of this simple equation is of exponential form. The numerical values in the equations are derived from field measurements.

Be aware that the application of this universal rule assumes that the radiation in question is in the form of a beam. For example, downwelling light in the atmosphere or ocean from the sun is described by a different attenuation coefficient that must take into account the fact that the scattered light is not removed from the system but can still contribute to the overall radiation. In the beam case, scattering removes light from the beam in an explicit way.

Another case to consider occurs when viewing a scene through a turbid atmosphere. The images are clearest only when you see the rays that come directly from the object, without any additional, scattered, light. If the atmosphere has only absorption and no scattering, the image remains highly resolved but dim. Under those conditions, Beer's law is completely accurate. The addition of a small amount of scattering in the scene also preserves the rule, since any light removed from the path that goes directly to your eye is lost forever. Moreover, in a condition of light scattering, rays from the sun and other light sources pass through the scene in front of you without getting scattered into your eye.

When multiple scattering is introduced, it can result in a violation of the rule, since some of the light removed from the most direct path may still reach your eye. In addition, rays that are passing through the scene can become scattered into your eye, increasing the brightness of the intervening atmosphere. In this case, the image is dimmed by absorption, and non-scene light is detected as well, leading to poor images.

Beam communication systems avoid the effects of multiple scattering to a large degree by having a narrow-field-of-view receiver that rejects almost all of the multiply scattered light, preserving Beer's law. A wide-angle receiver, such as the human eye, does not reject the scattered light, so Beer's law can be violated. The multiply scattered light appears as haze.

When using Beer's law, the user should verify that all units agree and be sure to use the correct bandpass when determining the scattering coefficient.

If changes of altitude are involved, be sure to note that α depends on altitude in the atmosphere. This adds some complexity to the calculation above.

Rarely can one number be used to represent total in-band transmission. Such simplifications are inevitable when a simple set of equations is used to model a complex system like the atmosphere.

The presence of an exponential attenuation term in transmission of the atmosphere is no surprise, since this mathematical form appears for a wide variety of media (including the bulk absorption of optical materials as in optical fibers). The ability to estimate attenuation in the presence of rain is convenient for many applications. The link in the atmosphere is subject to considerable limitation due to small particles, even if the rainfall rate is low.

The value of γ due to rain can be estimated in the following way:

$$\gamma \sim 3.92/V$$

V represents visual range (usually defined as contrast of 2 percent between target and object), and, for cases of Rayleigh scattering, γ varies as λ^{-4}.

Hudson[1] also shows that the effect of rain on visual range and scattering coefficient can be estimated from

$$\gamma = 1.2510^{-6} \frac{Z}{r^3}$$

where Z = rainfall rate in centimeters per second
         r = radius of the drop in centimeters

Alternatively, Burle[2] gives the scattering coefficient of rainfall as

$$\gamma = 0.248 f^{0.67}$$

where f = the rainfall rate in millimeters per hour

References

1. R. Hudson, *Infrared Systems Engineering*, John Wiley & Sons, New York, pp. 161–165, 1969.
2. *Burle Electro-Optics Handbook*, Burle Industries, Lancaster, PA, p. 87, 1974.

CROSS SECTION OF A RETRO-REFLECTOR

The cross section of a cube corner retro-reflector exposed to an approximately planar wave and viewed from the position of the laser is

$$\frac{D^4}{4\lambda^2}$$

where D = edge dimension of the cube in meters
λ = wavelength

Discussion

The diffraction-limited return beam half angle is $1.22\ \lambda/D$, so the beam fills an area of $\pi(1.22(\lambda/D))^2$ for a circular retro. This is the same as the solid angle of the beam emitted by the retro and is approximately $4(\lambda^2/\text{area})$. This result is the same for a retro with a square aperture. Cross section is defined as the inverse of the solid angle divided by the area. Thus,

$$\text{Cross section} = \left(4\frac{\lambda^2}{\text{area}^2}\right)^{-1} = \frac{D^4}{4\lambda^2}$$

This rule applies to "cube corner" or "corner cube" retro-reflectors only. Occasionally, such devices will have a slightly smaller cross section due to less-than-unity reflection and less-than-perfect tolerances on the angles of the mirrors.

This rule gives an immediate estimate of the detectability of an object equipped with a retro-reflector. It also allows the reflector to be sized so that detection at an appropriate range, and for a particular laser power, can be estimated. Additionally, targets (even noncooperative ones) frequently have structures that approximate cube corners, giving them a much larger signature than would otherwise be assumed.

Retro-reflectors are often used in tracking and pointing experiments to assure that the target is detected and the experiments can be carried out reliably. The reason is clear. By providing even the smallest retro-reflector, the target's signature is large and can be seen at great distances, even with laser systems of modest power.

An unfortunate term has crept into the literature concerning cube corners. The expression "corner cube" is much more common but is misleading. A retro-reflector can be formed from the corner of a cube over restricted angles. The term "corner cube" does not describe the geometry of any retro-reflector currently in use, but it is frequently used.

As an example, consider a cube of 0.1-m edge dimension exposed to visible (0.5 μ) light. The result is 100,000,000 m^2, 100 km^2, which is rather huge. Thus, the presence of a retro-reflector makes a target behave as if it is many orders of magnitude larger than its physical size.

FREE-SPACE LINK MARGINS

Weather condition	Required link margin, dB/km at 1550 nm
Urban haze	0.5
Typical rainfall	3
Heavy rainfall	6
Typical downpour, typical snow, light fog snow	10
White out snowfall or moderate fog	20
Heavy to very heavy fog	30 to >90

Discussion

Key to successful deployment of any free-space optical telecom link is its availability—the amount of time that the network is operating per unit of calendar time. In a rule presented in Chap. 6, it is asserted that network availability should be about 99.99 percent or better. This equals a downtime of less than a scant 9 s/day. Free-space telecom has the same equipment reliability and availability issues as any other network, but these systems also experience additional downtime due to weather. Atmospheric absorption, scatter, and scintillation all will decrease the signal-to-noise ratio and, if bad enough, will eliminate the ability for an optical telecom system to discriminate a "0" from a "1."

The above table gives some guidelines for the link margins that are suitable for various weather types. This is highly subjective, as all of these weather conditions, in the real world, can be "clumpy," both spatially and temporally. Such definitions are often in the eye of the beholder, which doesn't see in the 1550-nm bandpass, so the user must use these with an extreme grain of salt. However, they provide a good staring point.

The practitioner is encouraged to get the local weather statistics for the link and convolve these data with the above to determine the link margin needed for a given locale. Obviously, Adelaide and Tucson will need a lower margin for a given reliability than Seattle or Halifax, because there are far fewer occurrences in the former of weather that is deleterious to free-space transmission and that increases scatter (fog, snow, and rain). Additionally, link power and its resultant availability will often be seasonal; in many locations, bad weather tends to occur more frequently at certain times of the year.

The above link margins are for wavelengths of 1550 nm. The visible is slightly worse, and the longwave infrared (LWIR) is somewhat better. Carlson[1] gives the first five entries and data that also support the last entry, which is derived from the author's experience.

Reference

1. R. Carlson, "Reliability and Availability in Free Space Optical Systems," *Optics in Information Systems*, 12(2), SPIE, Bellingham, WA, Oct. 2001.

GAUSSIAN BEAM SIZE

In the far field, the full angular beam width of a Gaussian beam can be approximated by:

$$\varphi = \frac{2\lambda}{\pi w_o}$$

where φ = transmitter full angle beam width in radians
λ = optical wavelength in meters
w_o = Gaussian beam waist radius in meters

Discussion

This definition of beam width is based on the location of the $1/e$ points in the beam electric field. This is one of several common ways that beam spread is defined for Gaussian laser beams. Siegman describes several of them, including the more conservative 99 percent criterion in which the size of the beam is defined as the area that includes 99 percent of the beam's energy.

As pointed out above, rules of this type must be interpreted within the context of the definition of what constitutes the beam. Furthermore, the rule applies only in the far field, which is defined as propagation distance from the beam waist greater than $(\pi w_o^2)/\lambda$. This last distance is called the *Rayleigh range* and is equal to the distance from the waist at which the diverging beam has twice the area that it has at the waist. This is generally considered to be the range at which the beam makes the transition from near-field to far-field properties.

The beam width of a Gaussian beam is defined as the full width across the beam measured to the e^{-2} irradiance levels. Often one will encounter the beam width defined at the *full width, half maximum (FWHM)* points. These are also the –3 dB points. To convert a Gaussian beam profile specified at FWHM to the equivalent $1/e^2$ points, multiply it by ≈ 1.7.

References

1. G. Kamerman, "Laser Radar," in Vol. 6, *Active Electro-Optical Systems*, C. Fox, Ed., *The Infrared and Electro-Optical Systems Handbook*, J. Accetta and D. Shumaker, Ex. Eds., ERIM, Ann Arbor, MI, and SPIE, Bellingham, WA, pp. 15, 16, 1993.
2. A. Siegman, *Lasers*, University Science Books, Mill Valley, CA, p. 671, 1986.

LASER BEAM DIVERGENCE

A laser beam's full divergence angle is approximately the wavelength divided by the diameter of the transmitter aperture, or

$$\theta \approx \frac{\lambda}{d}$$

where θ = transmitter beam width (full angle) in radians
        λ = wavelength
        d = aperture diameter in the same units as the wavelength

Discussion

This rule derives directly from diffraction theory, particularly as applied to the subject of laser resonators.

This rule is widely used to estimate the size of a laser beam that has propagated through a vacuum and is also frequently used as a first estimate even in atmospheric applications. The rule works fine in environments in which scattering is small in comparison with absorption, since, in those cases, the beam shape is not affected.

This rule provides quick estimations of minimum beam divergence. Other rules in this chapter provide more detailed rules that might be applied if more accuracy is required.

If a beam is large in relationship to the receiving aperture, the power across the receiving aperture is nearly constant. The power at the receiving aperture is then a function of its placement along the far-field pattern, which can be described using a Bessel function. If the Bessel function is solved for the half-power points, the argument of the Bessel function

$$\frac{\pi d\theta}{2\lambda}$$

is equal to 1.62, and the beam divergence is equal to 1.03 λ/d. Note that the divergence is much smaller than the angular distance between the first zeros of the diffraction pattern (the Airy disk).

References

1. A. Siegman, *Lasers*, University Science Books, Mill Valley, CA, p. 56, 1986.
2. W. Pratt, *Laser Communication Systems*, John Wiley & Sons, New York, pp. 5–7, 1969.

LASER BEAM SCINTILLATION

Variance in the irradiance from a beam due to turbulence can be estimated as follows for the special case of a horizontal beam path.
  Variance in irradiance is defined as

$$\sigma = 4\sigma_t$$

where $\sigma_t = 0.31\ C_n^2\ k^{7/6}\ L^{11/6}$

  $C_n^2 =$ index of refraction structure constant discussed in another
      rule in this chapter (note that it has the odd units of
      meters$^{-2/3}$)

  $k = \dfrac{2\pi}{\lambda}$

  $L =$ propagation path length in the turbulent medium
  $\lambda =$ wavelength

Combining the equations above, we get

$$\sigma^2 = 1.24\ C_n^2\ k^{7/6}\ L^{11/6}$$

The standard deviation variation in irradiance is the square root of the
term on the right.

Discussion

This type of simplifying analysis of beam propagation in the atmosphere
has been spearheaded by both the military, who are interested in laser
beam propagation, and the astronomical community, who are very much
concerned with the disturbance that the atmosphere imposes on light collected by terrestrial telescopes. Of course, the latter group has little use
for analysis of horizontal propagation of light, but the underlying theory
that results in the equations above is related to the vertical views that
astronomers prefer.
  A basic assumption used in the development of the results for the horizontal beam case is that the value of C_n^2 is constant over the path. This assumption is generally not the case, since C_n^2 is the manifestation of
temperature variations in the atmosphere. However, this simplifying assumption is often used for horizontal beams and has been qualitatively
confirmed in many experiments. In the case that C_n^2 varies along the path,
a more complicated formalism must be used.
  The above equations apply to plane waves. Spherical waves can be characterized by the same analysis, except that the first equation uses 0.124 as
the multiplier rather than 0.31. In addition, there is a limit to the range of

atmospheric conditions over which the rule applies. The best estimate is that the above expressions can be used if σ_I is not larger than about 0.3.

Use this rule to make rapid assessments of the performance of laser beam systems and other light transmission equipment in the presence of atmospheric effects. The reference also provides the additional details necessary to deal with beam paths that are not horizontal. To do so requires that C_n^2 be known or estimated as a function of altitude. Other rules in the book illustrate how the computation of C_n^2 depends on altitude. Also, be wary about the size of the receiver that supports these statistical descriptions. What is shown here applies to cases in which the receiver is small as compared with the beam size, as measured by its lateral scintillation size.

The reader should note that a constant C_n^2 path will not give the same result, since sigma results from an integral over a spatial dimension. If you follow the constant C_n^2 path, it will be a longer distance path than the "horizontal" path.

That means that the integral will be different from the horizontal path unless the horizontal path and C_n^2 path coincide.

For example, if we use a typical C_n^2 value of 10^{-14} m$^{-2/3}$, a path length of 375 m, and a wavelength of 0.5 μ, we get an irradiance variance of about 0.12, which is equivalent to a standard deviation of about 35 percent variation in the intensity at the receiver.

Reference

1. J. Accetta, "Infrared Search and Track Systems," in *Passive Electro-Optical Systems*, Vol. 5, S. Campana, Ed., *The Infrared and Electro-Optical Systems Handbook*, J. Accetta and D. Shumaker, Ex. Eds., ERIM, Ann Arbor, MI, and SPIE, Bellingham, WA, p. 288, 1993.

LASER BEAM SPREAD

The angular spread of a beam along an atmospheric path (θ^2) is[1]

$$\theta^2 = \frac{1}{k^2 a^2} + \frac{1}{k^2 \rho^2}$$

where $k = \dfrac{2\pi}{\lambda}$
λ = wavelength
a = beam radius
ρ = transverse coherence distance (see below)

Discussion

The parameter ρ is the transverse coherence distance and is related to the effect that the turbulence of the atmosphere has on propagation of light.

$$\rho = \left(1.46 k^2 \int_0^L C_n^2(z) \left(\frac{z}{L}\right)^{5/3} dz \right)$$

where L = path length
C_n^2 = refractive index structure constant, which is further defined in the "C_n^2 Estimates" rule
z = a dummy path variable that covers the path over which the light travels

Although this equation looks daunting, remember that k is a constant, and C_n^2 for a given case may be as well. Even if C_n^2 changes along the path, one can compute the integral with a calculator if you keep your wits about you.

This rule comes from a combination of the analytic description of laser beam propagation in a vacuum along with a simplified assumption about the way other beam spreading effects, such as the atmosphere, add to the theoretical beam spread.

These estimates of beam size are as good as the quality of the estimates of C_n^2, since the description of beam spreading has been derived from the theory of laser resonators, which has been proven to be exact enough to predict all of the key features of laser radiation.

Finally, it should be noted that various authors use two different descriptions for ρ. One is for plane waves, and the other is for spherical waves. The expression for ρ above is the spherical wave case, which is the typical form used for laser beams propagating in the atmosphere. The plane wave case is typically used only for starlight propagating in the atmosphere.

This rule provides a quick and easy estimate of the size of a beam as a function of distance from the laser under the influence of the atmosphere. It is likely to be useful to those designing or modeling laser communications or LIDAR systems.

A special case pertains to a horizontal beam path, where

$$\rho = \left(1.46\frac{3}{8}k^2 C_n^2 L\right)^{-3/5}$$

for a horizontal beam.[2] This derivation is directly from the previous definition of ρ as presented above.

C_n^2 is defined in the C_n^2 Estimates rule (p. 99) and is about 1×10^{-14} (m$^{-2/3}$) near the ground and 1×10^{-18} (m$^{-2/3}$) at 20 km altitude.

Tyson[1] also reports that, for short propagation distances, a horizontal laser beam spreads to a waist size w_b, expressed as

$$(w_o^2 + 2.86\, C_n^2 k^{1/3} L^{8/3} w_o^{1/3})^{1/2}$$

and over longer distances, $L \gg \pi w_o^2/\lambda$.

$$w_b^2 = \frac{4L^2}{k^2 w_o^2} + 3.58\, C_n^2 L^3 w_o^{-1/3}$$

In these equations, w_o refers to the beam size as it exits the laser.

Laser beams are used for so many applications that the study of their beam spread in turbulence is of considerable use. Clearly, the simplest results are obtained when one can accurately assume that C_n^2 is constant. This is rarely the case, but the assumption is adequate for many applications. Considerable attention has been paid to this problem by a number of researchers. Recently, the emphasis has been in three areas: military applications, communications based on pulsed laser beams, and astronomical telescope systems in which a laser beam is used to create a synthetic star. The star is used to provide information on corrections that must occur to remove the effects of atmospheric turbulence.

References

1. R. Tyson, *Principles of Adaptive Optics*, Academic Press, San Diego, CA, p. 32, 1991.
2. R. Tyson and P. Ulrich, "Adaptive Optics," in Vol. 8, *Emerging Systems and Technologies*, S. Robinson, Ed., *The Infrared and Electro-Optical Systems Handbook*, J. Accetta and D. Shumaker, Ex. Eds., ERIM, Ann Arbor, MI, SPIE, Bellingham, WA, p. 180, 1993.

LASER BEAM SPREAD COMPARED WITH DIFFRACTION

A Gaussian spherical wave spreads considerably less than a plane wave diffracted by a circular aperture.

Discussion

The plane wave has an angular diameter due to diffraction of $2.44(\lambda/D)$ to contain 84 percent of the beam power, whereas a Gaussian spherical wave contains 86 percent of its power in an angular diameter of $2(\lambda/D)$. In both cases, D is defined as πw_o, where w_o is the radius of the waist (or smallest) size of the beam in the beam-forming optics.

This rule applies for diffraction-limited optics and does not include other aberrations. Furthermore, it does not include the effects of atmospheric scatter in applications in other than a vacuum.

Note the difference between this formulation and the more common diffractive properties of plane waves. That is, one should not assume that the typical diffraction formula,

$$2.44\frac{\lambda}{D}$$

which defines that beam divergence angular diameter, applies for lasers.

The specific results given above derive from a definition of the "size" of the beam. Since a Gaussian beam has an extent that is not well defined, some latitude must be accepted in the power numbers that are selected. For example, Siegman[2] points out examples in which the results vary, depending on how the beam radius is defined.

References

1. H. Weichel, *Laser System Design*, SPIE course notes, SPIE, Bellingham, WA, p. 72, 1988.
2. A. Siegman, *Lasers*, University Science Books, Mill Valley, CA, p. 672, 1986.

LASER BEAM WANDER VARIANCE

The variance (σ^2) in the position of a beam propagating in the atmosphere is

$$1.83 \, C_n^2 \, \lambda^{-1/6} \, L^{17/6}$$

where L = path length and the path is horizontal (the square root of this
number = standard deviation in the beam wander)
λ = the wavelength
L = distance the beam travels
C_n^2 = the atmospheric structure constant

Discussion

A whole generation of atmospheric scientists have worked on the problem of laser beam propagation in the atmosphere. Ultimately, all of the work derives from seminal analysis performed by the Russians, Rytov and Kolmogorov. Fried has also made important contributions to the theory. Military and astronomical scientists have extended the theory and have made considerable progress in demonstrating agreement between theory and experiment. The theory is too complex to repeat here. Fortunately, the effect on propagation can be expressed with relatively simple algebraic expressions such as the one shown above.

As with any rule related to the atmosphere, the details of the conditions really determine the propagation that will be observed. This result assumes that the value of C_n^2 along the path is constant and is of such a value that the turbulence effect falls into the category of "weak." This means that the variance in the beam intensity is less than about 0.3. Otherwise, the assumptions inherent in Kolmogorov's adaptation of Rytov's work no longer apply, and the results are flawed, although new research is ongoing for this regime.

Use of this rule allows us to define the required size of a receiver to encounter the bulk of a beam used for communications, tracking, or other pointing-sensitive applications.

The mathematics behind this analysis, which was first executed by Tatarski, is beyond the scope of this book. Suffice it to say that the result shown above is a rather substantial simplification of the real analysis that must be performed. For example, Wolfe and Zissis provide a more complete analysis and show how the beam wander is translated into motion of the centroid of the beam in the focal plane of a receiver.

Assuming that the nighttime value of C_n^2 is about $10^{-14} \text{ m}^{-2/3}$, a path length of 5000 m and a wavelength of 0.5 μ, σ is approximately 78 mm.

References

1. H. Weichel, *Laser System Design*, SPIE course notes, SPIE, Bellingham, WA, 1988.
2. W. Wolfe and G. Zissis, Eds., *The Infrared Handbook*, ERIM, Ann Arbor, MI, pp. 6–37, 1978.

PEAK INTENSITY OF A BEAM WITH INTERVENING ATMOSPHERE

Peak intensity (in W/m^2) of a beam going a distance L (in m) is

$$\frac{Pe^{-\varepsilon L}}{\pi L^2 (\sigma_L^2 + \sigma_B^2)}$$

where L = distance

σ_B = effect of blooming, in radians (likely to be essentially zero for any laser used for communications, but we include it for completeness)

ε = atmospheric extinction, meters^{-1}

P = beam power at the transmitter, watts

σ_L = combined effect of linear beam spread functions, equal to

$$\sqrt{\sigma_D^2 + \sigma_T^2 + \sigma_J^2}$$

σ_D = combined effect of diffraction and beam quality in radians

σ_T = effect of turbulence, in radians

σ_J = effect of jitter, in radians

Discussion

The beam intensity at a distance L is the result of the combined effect of beam spreading and atmospheric attenuation. The latter is contained in the exponential term in the numerator. It contains both the absorption, which removes energy from the beam, and scattering, which redirects the energy but removes it from the beam. The denominator simply describes the area over which the beam will be spread at distance L. It relies on several terms to describe the size of the beam, as described above.

Of course, the description of the beam shape is a simplification, particularly with respect to the atmospheric effects. The range of limitation really applies to the parts of the rule relating to atmospheric effects. The estimation of the impact of the atmosphere applies as long as the turbulence falls into the "light" (as opposed to "heavy") category as related in a previous rule.

This rule provides a quick and easy to program model of the intensity of a laser beam as a function of atmospheric conditions and the range to the target.

The diffraction effect is

$$\sigma_D^2 = \left(\frac{2BQ\lambda}{\pi D}\right)^2$$

where BQ = beam quality at the aperture

Note that when $BQ = 1$, we get the diffraction effect, which cannot be avoided.

When $D/r_0 < 3$, which applies for short paths,

$$\sigma_T^2 = 0.182\left(\frac{\sigma_D}{BQ}\right)^2\left(\frac{D}{r_0}\right)^2$$

where $r_0 =$ Fried's parameter

The value of r_0 varies from about 10 cm for a vertical path through the entire atmosphere to several meters for short horizontal paths.

When $D/r_0 < 3$,

$$\sigma_T^2 = \left(\frac{\sigma_D}{BQ}\right)^2\left[\left(\frac{D}{r_0}\right)^2 - 1.18\left(\frac{D}{r_0}\right)^{5/3}\right]$$

Also, note that the first equation reverts to

$$\frac{PA}{BQ^2\lambda^2L^2}$$

when there are no atmospheric effects.

Reference

1. R. Tyson and P. Ulrich, "Adaptive Optics," in Vol. 8, *Emerging Systems and Technologies*, S. Robinson, Ed., *The Infrared and Electro-Optical Systems Handbook*, J. Accetta and D. Shumaker, Ex. Eds., ERIM, Ann Arbor, MI and SPIE, Bellingham, WA, p. 198, 1993.

OPTICAL PERFORMANCE OF A TELESCOPE

A blur circle's angular diameter may be approximated by

$$\theta = \sqrt{\theta_D^2 + \theta_F^2 + \theta_A^2}$$

where θ_D = diffraction effect
θ_F = figure imperfection effect
θ_A = misalignment of optics effect

Discussion

A typical expression for θ_D is 3.69 λ/D for a system with a central obscuration radius of 0.4 of the aperture radius and $\theta_F = 16(x/D)$ where x is the optics figure. A typical value of x might be about $\lambda/3$ or $\lambda/4$. This value again assumes the obscuration mentioned above.

The effect of misalignment is approximately

$$\theta_A \approx \frac{mD\Delta}{k_1 EFL^2}$$

where D = diameter of the optical system
λ = wavelength of operation (or cutoff of a wide bandpass)
m = magnification of the telescope
Δ = lateral misalignment
k_1 = a constant
EFL = effective focal length of the telescope

This rule is based on an approximation of diffraction theory and general experience with telescopes and ray tracing. The real performance will depend on the optical design and some nonlinear features that are not represented in the equations. The expression for θ_D assumes a reflective telescope with a 40 percent obscuration and should be adjusted for other obscuration ratios. The rule provides a quick estimate of the blur size that can be achieved for a given level of optical quality and alignment effort.

It is important to understand that these imperfections in optics sum via root-sum-square (RSS). Having an element with exceptional figure does not provide a smaller spot if the alignment cannot be done properly. This gives a blur circle roughly equivalent to the Airy disk diameter (containing about 80 percent of the energy entering the telescope) for an obscuration ratio of secondary to primary of 40 percent.

Reference

1. NASA-Goddard Space Flight Center, *Advanced Scanners and Imaging Systems for Earth Observations*, U.S. Govt. Printing Office, Washington, DC, pp. 99–101, 1973.

POINTING OF A BEAM OF LIGHT

The probability, P_h, that a beam of angular radius θ_d will be pointed to within its radius (θ_d) is

$$\sigma = \frac{\theta_d}{2} \frac{1}{\sqrt{-2\ln(1-P_h)}}$$

where the pointing error is defined by a Gaussian with a standard deviation of σ.

Discussion

Note that the definition of the beam angular radius is a choice to be made by the user of this equation. For example, it could be the half-power point of a Gaussian laser beam or the $1/e$ point. The equation applies in either case, but the choice must take into account that the definition of beam radius determines the actual amount of power imposed on the target. Figure 5.1 illustrates the example. Note that the equation applies regardless of the distribution of energy in the beam. For example, if the beam has a Gaussian distribution of energy, θ_d could describe the $1/e$ points of the beam.

The probability that the beam is within a solid angle Ω is

$$P(\Omega) = \frac{\exp\left(-\frac{\theta}{2\sigma^2}\right)}{2\pi\sigma^2}$$

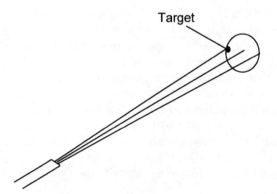

Figure 5.1 Pointing of a laser beam to within the beam radius.

The probability that the beam is pointed at the target within the angle θ is

$$P(\theta) = P(\Omega)\frac{d\Omega}{d\theta} = \theta\frac{\exp\left(-\dfrac{\theta}{2\sigma^2}\right)}{\sigma^2}$$

If we want to compute the probability that the beam is pointed to within one-half of the beam diameter, we integrate $P(\theta)$ over that interval,

$$P = \int\limits_0^{\theta_d/2} \theta\frac{\exp\left(-\dfrac{\theta}{2\sigma^2}\right)}{\sigma^2}\,d\theta$$

Of course, other measures of merit can be chosen as well, such as $1/10$ of the beam diameter. For the case of $\theta_d/2$, some manipulation results in

$$P = 1 - \exp\left(-\frac{\theta_d^2}{8\sigma^2}\right)$$

Solving for σ, we get the equation that appears at the beginning of this rule. The rule, as stated, applies to the illumination of a point target by a Gaussian beam. Larger targets are, of course, easier to hit. The size of the target is added to $\theta_d/2$. The analysis also assumes that there is no bias in the pointing of the beam.

This rule can be quite useful in defining the pointing necessary to illuminate a target or the sizing of the beam necessary to achieve the desired probability of hit.

If one performs calculations at the half-power point or $1/e$ points of a laser beam's intensity, one is building in a margin. Usually, the power will be 50 to 70 percent higher. If multiple "hits" or observations are allowed, then there is a comfortable built-in margin, as it is unlikely that random errors will result in several observations at the minimum points.

Suppose we have a beam that has a divergence, measured in half cone angle, of 10 mrad. What pointing is necessary to assure that the beam will encounter a point target with a probability of 0.99? This works out to be 1.67×10^{-3} radians. This result is consistent with the general conclusion that there is a high probability of pointing to within the beam radius, thus the pointing must be about $1/5$ of the radius.

Of course, the equation can be manipulated to calculate the probability of hit given the beam dimension.

SIGNAL-TO-NOISE REQUIREMENTS

An extremely high signal-to-noise ratio (SNR) is required at the receiver to reach the bit error rates that are desirable in telecom systems: 10^{-12} to 10^{-14}. For systems with forward error correction, it is common to have values in the range of 5 to 25 dB.

Discussion

A number of interesting and useful rules derive from Ref. 1. For example,

- A critical issue in any communication system is the coverage area that can be achieved. Beam spread of an optical system is a significant factor, and it scales with range squared.
- Beam shape must be chosen as a balance between power at the receiver (which is increased with a more tightly formed beam for a given laser power) and other factors (e.g., building sway, the quality of pointing that can be achieved, refraction by the atmosphere, and so on).
- Care must be taken to avoid interference in closely packed installations. This can be done by maintaining a minimum distance between installations or by using different wavelengths for different installations.
- As compared with local multipoint distribution system (LMDS), free-space optics (FSO) create an extremely narrow beam that significantly simplifies the problem of user separation. Since atmospheric attenuation is a function of visibility, a weather database is necessary. This database can be used to set the link distances at levels that ensure a reliable link.
- Optical transmitter launch power must be set to optimize link margin and BER for the wavelength, distance, optical receiver sensitivity, and worst-case atmospheric conditions.
- The FCC has chosen not to regulate frequencies greater than 300 GHz (less than approximately 10,000 nm in wavelength). Since FSO systems all fall well above 300 GHz, there is no need for any spectral licensing. Furthermore, because the footprint of the beam is so small, the interference problems that can arise in unlicensed radio pose no problem for FSO.

Reference

1. B. Willson et al., "LMDS versus Free Space Optical Networks," *Proceedings of the National Fiber Optic Engineers Conference*, 2001.

6

Network Considerations

Everyone seems to loosely use the term *network* to describe the unseen hardware and software that almost seamlessly and sometimes almost magically transports our data and voice—and soon our videos. The word *network* is even shortened to *net* as common vernacular for the Internet and can be used as either a verb or noun. Yet it is difficult to define exactly what the *net* is. To most users, it is almost ethereal. Its hardware is typically invisible to the user (ignoring the proliferation of those ugly cell towers) and requires little or no control by the user, and users barely think of the myriad of fibers, lasers, and switches that it takes to download that swimsuit photo or talk to Mom.

A telecommunications network can be loosely defined as all of the hardware and software required for transmitting, routing, and delivering information, including all of the lines and interconnects. However, sometimes the hardware and software used to generate, collect, or display the information is not included in the concept of the "net." As such, networks owned by various companies need to seamlessly interact and connect to other networks.

Network architecture is an eclectic mix of art, science, and gut feelings. In addition, a vast amount of time is spent (usually by professional organizations) to create standards that enable the net. There is always a desire for the ability to engineer a network and achieve the maximum performance at minimal cost. However, even this simplistic metric is difficult to gauge, as the definition of maximum performance can change over the time that it takes to deploy a network. The cost calculations for the maintenance, the future software required to interface with another company's network, and reliability can be changing variables.

The simplest communication network design started thousands of years ago for armies to communicate with each other and the home control. Perhaps the Greeks became the first optical information network architects when they planned out upon which mountains lanterns would be posted to assure an unbroken stream of information across a linear network. The information was only tens of bits per hour, but better than nothing, and it did travel between nodes at the speed of light.

The first telephone networks began to appear in the latter part of the nineteenth century. They depended on having every phone connected to a local central office. The operator functioned as a human switch and could connect you via patch cords to anyone else in your town. Shortly, after that, one or a few long-distance connections were added to each operator's station, where she (it was almost always women, for some reason) would patch you into the long distance operator. By the 1920s, these human-operated mechanical switches became unwieldy and laborious, resulting in the development of all-mechanical switches and ingenious efficiency titivations. Deployment was slow, with human operators providing at least some switching services into the 1980s in some rural parts of the world. In 1917, A. Erlang began the theoretical development of the subject of network scheduling and data rates. Shannon published landmark papers in information theory in the late 1940s and early 1950s.

Technology always seems to be embraced and expanded by the military; telecom networks were certainly no exception. Both WWI and WWII saw major developments in network technology as well as network philosophy by the military in its search for efficient and secure communications. Teams were constructed to devise and design the communications network to support operations, especially dealing with cyphers and codes. As military forces are mobile, most of the effort was with wireline networks.

In the late 1940s, network design became a scientific and engineering subject with the theoretical efforts of Claude Shannon and others. Telephone networks continued to expand in complexity, and satellites were added in the 1960s. The Internet (originally called the ARPAnet) was invented in the 1970s by the U.S. Department of Defense to link research computers. Significant advancements in data transfer and the paperless office concept were being achieved at Xerox's Palo Alto Research Center. The synchronous optical network (SONET) system was developed in the 1970s as a redundant ring network, and the first fiber telephone networks appeared in the 1980s and were advertised as being so quiet that your could "hear a pin drop."

However, the explosive growth of bandwidth to supply data communications and the Internet of the 1990s saw billions of dollars pumped into network architecture and design as well as into funding new innovative optical components. The World Wide Web was pioneered at CERN (in France and

Switzerland) in the early 1990s and became the ubiquitous Internet path by the end of the decade.

Since its inception in the mid-1970s, Internet traffic has approximately doubled every year. For a short time in the mid-1990s, bandwidth demand was growing at about 100 percent every six months. However, the willingness to pay for this increased bandwidth did not grow at the same rate as the bandwidth demand. The price elasticity of bandwidth resulted in users demanding ever more bandwidth at ever lower costs. Many thought that income and profits would also double or even grow faster each year, and for a short time the stock market reflected these expectations.

The bandwidth bubble transferred to a bubble in corporate market capitalization, resulting in many billionaires and almost uncountable millionaires—until the bubble burst in March of 2000. Recent post-bubble efforts have concentrated on more realistic growth in network bandwidth of 10 to 30 percent per year with revenue growth of 10 percent or less (as the price of bandwidth continues to fall).

Post-bubble technical efforts concentrate on replacing electrical network elements with all-optical components and subnets. Another area of interest focuses on packing ever more wavelengths (components supporting nearly 1000 laser lines per fiber have been brought to market) of light into fibers as well as moving from the old ubiquitous 2.5 Gb/s/fiber data stream to 40 or 80 Gb/s.

Typically, networks are divided into four species:

- Ultra-long-haul
- Long-haul
- Metro
- Enterprise

The ultra-long-haul typically support the backbone and can send data thousands of kilometers (and often many hundreds or thousands of kilometers) before a switch is encountered. They include under-ocean transport. The long-haul networks provide data transmission from 100 to 1500 km, typically between cities in the U.S. and Asia and between countries in Europe. The metro systems are the inter-rings or point-to-point subnetworks that typically surround a city; they contain numerous switches and serve millions of points of presence (POPs), but they generally do not transmit data farther than across town (up to ≈200 km). The enterprise networks are even smaller in geographic area and support a bandwidth-intensive "campus," be it a company, building, or university, and they frequently incorporate firewalls and monitored internal transmissions. Networks tend to be classified as a ring (such as SONET), mesh, or linear. Rings and meshes allow redundancy, and many feel that meshes offer increased performance on all optical implementations. The goal of moving to all optical connections, switches, and routers is an obvious one with the deployment of optical telecom. In the 1990s, when an optical path was deployed, the optical signals had to be converted to electrical signals for some amplification, all routing, and all switching, then converted back to optical for transmission across the next link. The detection process resulted in optical-to-electrical conversion and more complexity. The pressing need to control cost demands that streamlined all-optical processes take over these complex interfaces. The obvious inefficiencies resulted in increased rack space, increased electrical consumption, and limited data rates. Mike Arden[1] points out that another key benefit to all-optical devices is that they are bit-rate agnostic. That is, a 2.5-Gb/s wavelength and a 10-Gb/s one can pass through the same all-optical component without having to reconfigure the component. This saves time and money while enabling reconfigurable networks. With ever increasing bandwidth to be required by streaming video, wireless Internet, and more user-friendly services, the migration to more optical network elements and fewer electrical ones is in progress. It was realized, way back in the 1990s, that electrical components were reaching their maximum bandwidth, while all-optical components can operate at a much higher bandwidth and allow additional network growth.

Whether squeezing more bandwidth from legacy networks or developing completely new networks, the practitioner requires some basic background information as well as an appreciation for what has worked in the past and what has not. Hopefully, the rules in this chapter will provide a start down this path. The included eclectic rules will help the reader understand this process. Included in this chapter are some stalwart concepts such as Shannon's rule, the Erlang definition, and the law of locality. Along with these are some definitions of the amount of data required to perform a task and just what a terabit means. These are included to remind the designer of the power of this work and of this information-trading technology. Some rules are specific to hypernets, fiber lengths, and growth. Several of the rules deal with reliability or maintainability, as this is a critical subject that must be addressed from network architecture through design, deployment, and (obviously) maintenance.

Reference

1. Private communications with M. Arden, 2002.

APPLICATION BANDWIDTH

The following applications require approximately the stated bandwidth per user:

Application	Bandwidth in Mb/s
HDTV (uncompressed)	1200
1280 × 960 pixel, 60-Hz video	590
640 × 480 pixel video, 3-Hz video	74
Compressed high-definition video session	19–20
320 × 240 pixel video, 15-Hz video	9
Standard digital video sessions	6
NTSC video	6
Compressed VHS video	3
Internet appliances	2
Internet gaming session	2
CD-quality audio	1.2
Video conferencing sessions	1
Broadband web surfing session	0.3
High-quality audio sessions	0.125
Voice	0.064

Discussion

There have been a plethora of projections of the increased bandwidth per year. The authors of this book have kept away from such forecasts, as they have all proven to be too optimistic for the early part of the twenty-first century (reference our investments). However, it is clear that bandwidth will, in fact, continue to grow faster than world population grows, as more people get constant and reliable access to the Internet (due to dropping prices for computers and access as well as improving high-bandwidth services) and as the Internet begins to serve more data-intensive applications such as video.

For the foreseeable future, the bandwidth growth for telecom networks clearly will be data driven, as voice takes a backseat. Data is expected to account for 90 percent of all traffic by 2004.[1]

Some of the video applications mentioned above assume no compression and an 8-bit format, although most displays are more like 6-bit. A 320 × 240 pixel image corresponds to an acceptable low-quality (less than VHS) instructional video. A 640 × 480 pixel image is approximately SVHS, and the noninterlaced 1280 × 960 is high-quality (but not movie-quality) video. Obviously, streaming video is the next big bandwidth driver.

The above can be used as a guide to what applications require, and the user can then estimate the mix as a function of time for users, and thus formulate a more realistic and accurate projection of bandwidth. Lund[2] points out that a 7-ft node rack can service 3500 subscribers. If a future application mixture is uniformly distributed between the first 10 applications above, the average user would require about 70 Mb, and 3500 users would require 247 Gb/s.

Figures 6.1 and 6.2 show graphically the required bandwidth for various applications and the total bit rate for time division multiplexed (TDM) networks and various DWDM wavelength division multiplexed networks. These two diverse graphics can almost be used as a nomograph, where one looks at the first, finds the required bandwidth, multiplies by the number of expected users to determine total bandwidth required, then goes to the Nakazawa plot and finds the type of architectures that can support this data rate. The arrow in Fig. 6.1 indicates high-bandwidth future applications. The provider that can economically address these will capture a large market share.

References

1. M. Simcoe, "A Multi-service Voice, Data and Video Network Enabled by Optical Ethernet," *Proceedings of the National Fiber Optic Engineers Conference,* 2001.
2. B. Lund, "Fiber-to-the-Home Network Architecture: A Comparison of PON and Point-to-Point Optical Access Networks," *Proceedings of the National Fiber Optic Engineers Conference,* 2001.
3. M. Nakazawa, "Solitons for Breaking Barriers to Terabit/Second WDM and OTDM Transmission in the Next Millennium," *IEEE Journal on Selected Topics in Quantum Electronics,* 6(6), p. 1133, November 2000.
4. C. Lu, *The Race for Bandwidth,* Microsoft Press, Redmond, WA, p. 155, 1998.

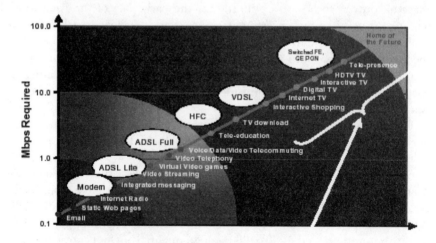

Figure 6.1   Application flow vs. data flow requirements (from Ref. 1).

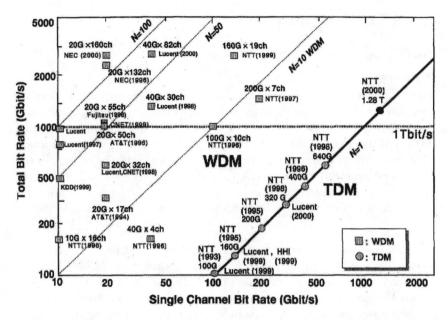

Figure 6.2   Total bit rate vs. single-channel bit rate (from Ref. 3, copyright © 2000, IEEE).

DATA TRAFFIC GROWTH

The amount of data traffic doubles every year.

Discussion

There was a time when it was often claimed that Internet traffic doubles ever 100 days or so. This rapid growth was, in fact, experienced at certain times during 1995 and 1996, but it has simply not held true over the long term, and there have been many periods of much slower growth. Although there are month-to-month and year-to-year variations in the change in the amount of data traffic, it seems that, on the long-term average, it has only approximately doubled every year and can be expected to continue this growth or less for the foreseeable future. Rocha backs up these claims as follows:

> Internet traffic is growing at a rate of 100 percent per year. At this rate, data traffic will surpass voice traffic around 2002. Although still considered to be a very fast rate of growth (100 percent), this rate is much slower than the frequently heard claims of a doubling in traffic every three or four months from Internet usage, claims which were based on the 1995-1996 time period, but far in excess of historical growth patterns.[1]

This growth reflects complicated interactions with society, economics, and people's ability to adopt change, as well as purely technological issues. The trend through most of 2001 was that the number of web sites actually dropped, but this is expected to be a temporary phenomenon.

Coffman and Odlyzko[2] did a comprehensive study of the data traffic on the Internet since ARPAnet was formed in 1969, and their detailed paper appears on the cited web page. Incidentally, the authors of Ref. 2 also point out that, sometime in 1997, the amount of operating disk space exceeded the amount of data that had ever been generated by humankind. Obviously, there is much duplication of data in copied files and operating systems. However, as world disk capacity and the data stored on it increases vastly, the implication is that some new data will never be seen by human eyes. A direct cause for this increase has been the dramatic drop in the cost of disks. At the time of this writing, optical CDs cost less than $0.50 per gigabyte, compared to several dollars per gigabyte just a few years ago.

References

1. M. Rocha, "Summary of Internet Growth: Is There a Moore's Law for Data Traffic?" 2002, http://www.pegasus.rutgers.edu/~youssefm/teaching/cis604/summaries/paper3.html.
2. K. Coffman and A. Odlyzko, "Internet Growth: Is there a 'Moore's Law' for data traffic?" 2001, white paper from http://www.research.att.com/~amo/doc/networks.html.

DEFINITION OF AN ERLANG

Erlang = (message or packet arrival rate) × (average holding time)

We define $B(a,n)$ as the probability of blocking (i.e., the probability that a packet or call will find all circuits busy and be forced to join a queue) given n circuits and a erlangs. This is computed as

$$B(a,n) = \frac{\left(\dfrac{a^n}{n!}\right)}{\displaystyle\sum_{i=0}^{n} \dfrac{a^i}{i!}}$$

and

$$C(a,n) = \frac{B(a,n)}{1 - \dfrac{a}{n}(1 - B(a,n))}$$

where $C(a,n)$ = probability of blocking

The average delay time is

$$T = \frac{L}{R} + \frac{C(a,n)L}{\left(1 - \dfrac{a}{n}\right)nR}$$

where T = average delay time
   L = message length in bits
   R = line rate in bits per second

The average queue length in packets is given by

$$q = \frac{\dfrac{a}{n}C(a,n)}{1 - \dfrac{a}{n}}$$

where q = average number of packets in a queue

Discussion

The erlang (generally used without capitalization, although it is named for a person) is a dimensionless telephony unit that is still employed in network management of traffic flow. Conceptually, it is equal to a single user using a single link 100 percent of the time. Thus, statistically, 20 users who each use the resource 5 percent of the time also represent 1 erlang. The erlang is dimensionless, as when calculated properly, all the units cancel. It is named for Agner K. Erlang [1878–1929], a Danish mathematician who studied queuing theory and published a series of related equations in 1917.

The $B(a,n)$ equation is often called the *Erlang B*, and the $C(a,n)$ equation is often referred to as the *Erlang C*.

Stern and Bala[2] point out that, when N/C is 5, the Erlang B equation is a fairly good approximation of a finite population case. There are numerous simplifications of the Erlang B equation, with the most useful generally being the Markovian and Engset models, which are described by Stern and Bala in the same work.

The probability of blocking by a given number of circuits with a load of a Erlang B is often a difficult equation to solve due to the factorials. Modern telecom networks tend to have hundreds or thousands of individual "circuits" or links, and even large computers will stall when dealing with a factorial of 1000 and numbers raised to the power of 1000. To allow computation, Pecar, O'Connor, and Garbin[1] suggest using the following recursive form of the equation in a computer loop:

$$B(a, k) = \frac{aB(a, k-1)}{k + [aB(a, k-1)]}$$

where $B(a,k)$ = probability of blocking provided by k circuit links under a load of a erlangs
   a = number of erlangs
   k = number of circuits

Each time this equation is run in a "loop" and k is incremented, an estimation closer to the closed form results. Pecar also points out that this recursive form is useful for quick estimations of the number of circuits required to achieve a given grade of service. One keeps running the equation and increasing the number of circuits until the blocking is low enough.

References

1. J. Pecar, R. O'Connor, and D. Garbin, *Telecommunications Fact Book*, McGraw-Hill, New York, pp. 123–132, 1993.
2. T. Stern and K. Bala, *Multiwavelength Optical Networks*, Prentice Hall, Upper Saddle River, NJ, pp. 335–337, 421–423, 2000.

GROWTH FUNCTION

$$G(t) = \frac{\beta}{1 + \alpha\exp^{-mt}}$$

where $G(t)$ = growth rate as a function of time
   α = initial rate
   β = expected value at the end of the forecast
   m = a constant
   t = time scale (any time scale that you want as long as you are consistent with the forecasts in α and β)

Discussion

The above exponential equation is well known for forecasting network growth, be it the bandwidth, number of users, nodes, or other factors. The exponential nature seems to hold true, backed by numerous empirical studies, even considering the downturn experienced at the turn of the century. One can theorize that the exponential form is due to the interconnectivity of a network and, at least partially, results from Metcalfe's law. Growth in telecommunications results from either more users or the employment of more bandwidth by existing users. In the former case, "more users" means that there are more people with whom we may communicate and exchange data. Thus, a doubling in the number of users usually means that demand grows more than a factor of two, as the previous users are also communicating with the new users, and everyone is using the network more. A similar, albeit less powerful, argument can be made for the same number of users increasing their bandwidth use. It became popular in the 1990s to send e-mail jokes. Generally, a single person would send it to many users, who would then forward it on to even more users. Frequently, people received the jokes more than once. In the latter 1990s, it became commonplace to send joke images. These required much more bandwidth but seemed to be sent to just as many people, with just as many repeats. The images also represented a quicker and easier joke and tended to be sent to even more people, although all of this required more bandwidth. Because of the extra increase in distribution, and more images being sent, the required bandwidth was more than just the growth in bandwidth by the change from printed jokes to images. Also, increased bandwidth capability usually results in more users being connected to the network. An analogy with highways can be made. Bigger and better-designed roads attract more cars and can be assumed to encourage more driving and more traffic jams.

   This growth rate is quite sensitive to all inputs, especially α, β, and m, in the above equation. α is the utilization rate at time zero, and β is the final value. The equation predicts the required values between the two. The constant m is critical, as it determines the rates of the exponential growth and must be carefully chosen.

Reference

1. M. Esfandiari, I. Hsu, and M. Sloboda, "Network Efficiencies of SONET Systems with ATM Functionality: Case Study of an Infrastructure Network," *Proceedings of the National Fiber Optic Engineers Conference,* 1997.

LAW OF LOCALITY

Network traffic is about 60–80 percent local, 90–95 percent continental, and only a few percent intercontinental.

Discussion

Generally, telecommunication network traffic has contained more local than long distance content. "This has been true historically, because you e-mail people that you know, and you tend to know people amongst whom you live," writes Gilder.[1] Additionally (with the possible exception of international salesmen), most people's work e-mail and telephone calls are to other members of the same organization.

In the mid and late 1990s, the Internet (which drove data bandwidth) was largely an American phenomenon, and the majority of data traffic was contained within the U.S.A. However, that will change as the Internet becomes more global in the new millennium. The globalization is underscored by Gilder, who asserts that "between mid-1997 and early 2000, the North American share of world-wide internet users dropped from over 80 percent to 55 percent."[1]

Most traffic being local, large-capacity enterprise and metro bandwidths are needed. While this is true, bandwidth is still required to be large for the long-haul (LH) and ultra-long-haul (ULH) networks as the number of users increases (even if their percentages are less, as indicated by above). The above point notwithstanding, there is some empirical evidence that existing optical networks already reflect the law of locality. Barry and Cao[2] indicate that existing ULH networks (3000 km) typically have 25 to 50 percent of the capacity of LH systems (1000 km). Incidentally, in a separate discussion, they point out that this is compounded by wavelength blocking necessary for DWDM networks, with studies showing a 30 percent decrease due to this optical blocking.

Additionally, Rocha claims, "Data traffic will grow in line with supply and demand as the central guiding point, and computer and network architectures would be strongly affected, as most data would stay local. In addition, 'new' technologies such as the Napster program will have a significant impact on data transmission."[3]

However, the cause and effect concerning this issue can be cloudy. Arden[4] gives a valid counterpoint: "Ultra-long haul (ULH) systems were designed to go very long distances. In the past, to achieve these distances, these networks compromised on capacity. Modern ULH systems are correcting this and frequently have the same or higher capacities as their long haul counterparts. Also, long-haul systems used to reach cross-continental distances have as high a capacity as they do with shorter distances. It tends to be the equipment employed rather than the distances that compromises capacity."

References

1. G. Gilder, *Telecosm*, The Free Press, New York, pp. 189–190, 2000.
2. R. Barry and Y. Cao, "The All Optical Network: When and How," *Proceedings of the National Fiber Optic Engineers Conference*, p. 124, 2001.
3. M. Rocha, "Summary of Internet Growth: Is There a Moore's Law for Data Traffic?" 2002, http://www.pegasus.rutgers.edu/~youssefm/teaching/cis604/summaries/paper3.html.
4. Private communications with M. Arden, 2002.

MAXIMUM NUMBER OF ERBIUM-DOPED FIBER AMPLIFIERS

The maximum number of erbium-doped fiber amplifiers (EDFAs) in a fiber chain is about four to six.

Discussion

The rule is based on the following rationales:

1. About 80 km exists between each in-line EDFA, because this is the approximate distance at which the signal needs to be amplified.

2. One booster is used after the transmitter.

3. One preamplifier is used before the receiver.

4. Approximately 400 km is used before an amplified spontaneous emission (ASE) has approached the signal (resulting in a loss of optical signal-to-noise ratio [OSNR]) and regeneration needs to be used.

An EDFA amplifies all the wavelengths and modulated as well as unmodulated light. Thus, every time it is used, the noise floor from stimulated emissions rises. Since the amplification actually adds power to each band (rather than multiplying it), the signal-to-noise ratio is decreased at each amplification. EDFAs also work only on the C and L bands and are typically pumped with a 980- or 1480-nm laser to excite the erbium electrons. About 100 m of fiber is needed for a 30-dB gain, but the gain curve doesn't have a flat distribution, so a filter is usually included to ensure equal gains across the C and L bands.

For example, assume that the modulated power was 0.5 mW, and the noise from stimulated emission was 0.01 mW. The signal-to-noise ratio is 0.5/0.01 or 50. If an EDFA adds a 0.5 mW to both the modulated signal and the noise, then the modulated signal becomes 1 mW, and the noise becomes 0.501 mW, and the SNR is reduced to 2. After many amplifications, even if the total power is high, the optical signal-to-noise ratio becomes too low. This typically occurs after four to six amplifications.

Another reason to limit the number of chained EDFAs is the nonuniform nature of the gain. Generally, the gain peaks at 1555 nm and falls off on each side, and it is a function of the inversion of Er^{+3}. When a large number of EDFAs are cascaded, the sloped of the gain becomes multiplied and sharp, as indicated in Fig. 6.3. This results is too little gain-bandwidth for a system. To help alleviate this effect, a gain flattening device often is used, such as a Mach–Zehnder or a long-period grating filter.

Reference

1. A. Willner and Y. Xie, "Wavelength Domain Multiplexed (WDM) Fiber-optic Communications Networks," in *Handbook of Optics*, Vol. 4., M. Bass, Ed., McGraw-Hill, New York, pp. 13–19, 2001.

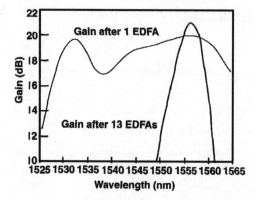

Figure 6.3   Gain vs. wavelength (from Ref. 1).

MAXIMUM SPAN LENGTH

The maximum optical span length is close to

$$L_s = 6P_{dBm} + 250$$

where L_s = maximum fiber span length in kilometers
  P_{dBm} = launch power in dBm

Discussion

Repeaters, amplifiers, and other associated hardware are real cost drivers and data limiters for long-haul (LH) and ultra-long-haul (ULH) systems. A longer span can result in lower cost and a better network. It is useful to understand the maximum length of a typical span for quick calculations at meetings and planning sessions.

The above rule is based on rounding up terms derived by the source,[1] assuming that the single-point nonamplified span length must be less than

$$\frac{(\text{Launch power in dBm}) - (\text{receiver sensitivity in dBm})}{\text{Fiber loss in dB per km}}$$

It assumes a zero-dispersion fiber and neglects nonlinear effects and margins, which is acceptable for a rule of thumb concerning the maximum span. The source calculated this for a G654 fiber with a loss of 0.175 dB/km and a sensitive optical receiver of –42 dBm @ 2.5 GHz and obtained

$$L_s = 5.7P_{dBm} + 240$$

Note that this rule is not valid for solitons.

Reference

1. A. Mitchell, Z. Zhu, and X. Zhang, "Advances in Repeaterless Systems," *Proceedings of the National Fiber Optic Engineers Conference,* 1999.

CHANNEL POWER

The maximum power that a wavelength channel should have is <17 dBm; expressed in milliwatts, it should be less than

$$\text{Individual channel power} < \frac{60 \text{ mW}}{n_\lambda}$$

where n_λ = number of DWDM wavelengths

Alternate Version

Generally, all the power should add up to no more than 17 dBm, and the average power is reduced by 2 dBm for every wavelength added.

Discussion

Bright laser light can cause injury to human eyes, so the amount of light pumped into a fiber should be within safe levels to allow easy repair and deployment and mitigate any safety hazard. The issue of eye safety is addressed in detail in a rule in Chap. 13 (p. 329). Fortunately, the human eye is not sensitive at the C and L bands, as the light is not focused on the retina. This results in much higher safe light levels compared with visible light. Nevertheless, parts of the cornea can be burned by laser light with long exposures to bright light; 62 mW is considered to be the approximate safe exposure level at the wavelengths of the C and L bands (this is approximately an order of magnitude less than the fluence demonstrated to cause temporary damage).

A level of 62 mW corresponds to about 18 dBm. Anderson and Jones[1] suggest a 1-dBm margin to accommodate multimode fibers and the extra light from amplified spontaneous emission of the fiber amplifiers. Moreover, optical spectrum analyzers and amplifiers are often specified to handle only 17 to 20 dBm of total power, as are many other network elements, so this level also represents a safe level for equipment.

References

1. G. Anderson and W. Jones, "Safety Considerations and Their Impact upon WDM System Deployment," *Proceedings of the National Fiber Optic Engineers Conference*, 1997.
2. ANSI document Z136.1, Laser Safety Standards, American National Standards Institute, New York, 2000.

MESH REDUNDANCY

The redundancy of a large network can be estimated as

$$R \cong \frac{2}{D-2}$$

where R = redundancy, defined as the total spare working capacity
D = average degree of nodes in a network (e.g., $D = 2S/N$, where S = number of spans in a network, and N = number of nodes in a network)

Discussion

The above assumes large networks with a high degree of connectivity such that the number of nodes is much greater than two, and the number of degrees of nodes is larger than two. Capacity is defined as the number of wavelength segments available for use in the network.

A more exact version of this equation is also given by the source[1] as

$$R = \frac{C_s}{C_w} = \frac{N-1}{S-(N-1)}$$

where C_s = total spare capacity in a network enumerated as the number of distinct spare wavelength segments
C_w = total working capacity in a network enumerated as the number of distinct working wavelength segments
S = number of spans in the network
N = number of individual nodes in the network

The average degree of nodes can be described as equaling $2S/N$. This allows simplification of the above equation to

$$R = \left(2 - \frac{2}{N}\right)\left(\frac{1}{D-\left(2-\frac{2}{N}\right)}\right)$$

The equation immediately above reduces to the rule for $N \gg 2$ and $D > 2$. Iraschko notes,

...without wavelength conversion, the redundancy of a large span restorable mesh network with $N \gg 2$ and $D > 2$ approximately doubles. Therefore, unlike ring survivable networks, the penalty of not performing wavelength conversion in mesh survivable networks can be severe. Without wavelength conversion, a mesh survivable network may require twice as many wavelength segments as a mesh network with wavelength conversion.[1]

Reference

1. R. Iraschko, "An Analysis of the Benefits of Wavelength Conversion in Ring and Mesh Survivable Networks," *Proceedings of the National Fiber Optic Engineers Conference*, 2000.

NETWORK AVAILABILITY

Network availability should be about 99.99 percent or better.

Discussion

This means that the network should be available for subscriber use all but 53 min or so per year. "Four nines of availability" is not impossible to achieve with backup systems and quick maintenance procedures. Obviously, these numbers do not include *force-majeure* situations such as massive terrorist attacks, wars, or localized outages during major (e.g., category 3 or higher) hurricanes. Generally, the failure rates do not include known periodic maintenance, such as replacing a component that is expected to wear out. Careful planning of the maintenance approach can reduce the chance that the service operation will affect customers. Maintenance that does not interrupt subscriber service is not included in downtime, but is included if service is interrupted. Whether noninterrupted, nonscheduled maintenance (e.g., an internal monitor sends an alarm to replace a transmitter in the next month because a laser's power is steadily decreasing) should be included is a subject of disagreement. This could occur if an internal monitor sends an alarm to replace a transmitter in the next month because a laser's power is steadily decreasing.

As networks get larger and more complex, a stickier issue can arise concerning degraded performance. For example, a subscriber paying for a 1 Mb/s DSL connect line might still find his DSL working, but only at a paltry 384 kb/s. Network providers and service providers have long taken advantage of their customers by supplying decreased performance, hoping that customers won't notice—or if they notice, won't complain. The authors would like to contend that such reduced performance should be considered a failure if it exceeds one percent of the time per year.

Availability is closely related to both the failure rate and the time it takes to repair them. There are several versions of availability equations to calculate availability, which can be found in Telcordia and military specifications. Below are two examples.

$$A_i = \frac{MTBF}{MTBF + MTTR}$$

where A_i = inherent availability (a number less than 1; this is the theoretical limit of the fraction of time that the system is available for use)
   $MTBF$ = mean time between failure (including corrective actions)
   $MTTR$ = mean time to repair

$$A_o = \frac{MTBM + RT}{MTBM + RT + MDT}$$

where A_o = operational availability (a number that represents the practical limit on the availability of the system)

$MTBM$ = mean time between maintenance (or preventive and corrective actions based on MTBF)

RT = component average ready time per cycle (available but not operating time per time period)

MDT = mean down time = mean active corrective time / (preventive maintenance time + mean logistic time + mean administrative time)

NETWORK FIBER LENGTH

The required fiber length can be estimated from

$$\text{Fiber length} = PDN$$

where P = a redundancy constant, usually between 1 and 4
D = distance
N = number of required fiber links (N end switches)

Discussion

The required length of fiber for a network is much more than the length of cable. This is due to redundancy. The above should be carefully considered when specifying a cable, the number of fibers, or the fiber cost, especially if it is an expensive or special fiber.

The above assumes that all the links have the same length. While obviously not universally true, this is a good assumption to begin with if you don't have all of the details and merely need to estimate the kilometers of fiber needed.

The redundancy constant P is important and interesting. The engineer should beware that this can greatly increase the amount of fiber needed. Lisle states,

> . . . where P = 2 if the OC-N (optical channel-n) interface is unprotected or 4 if 1+1 line APS (automatic protection switching) is used. For example, if 8 ATM CPE (customer premise equipment) devices are directly connected using 1+1 APS to a single ATM switch that is 2 miles away, the total required fiber length is 64 miles.[1]

The above values for P assume a SONET/SDH; an unprotected DWDM network would have a P of 1 (see Fig. 6.4). Perhaps the only case for a P of 2 for a DWDM is in a point-to-point system such as shown in the figure.

Reference

1. S. Lisle, "A survey of ATM Network Topologies," *Proceedings of the National Fiber Optic Engineers Conference*, 1997.

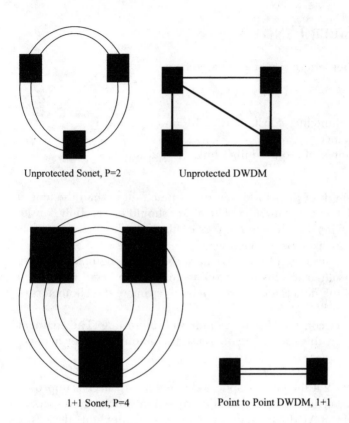

Unprotected Sonet, P=2   Unprotected DWDM

1+1 Sonet, P=4   Point to Point DWDM, 1+1

Figure 6.4   Network architectures.

NETWORK GUIDELINES

Gopalakrishnan suggests that special attention be paid to the following attributes from the beginning of network design:

- The ability to leverage existing infrastructure
- Bandwidth control and service provisioning
- Traffic engineering
- Real-time accounting
- Compatibility

Discussion

Although these were originally listed for a service creation platform for metropolitan networks, the basic premise is that these should be considered for all types of networks, and a checklist should be established. Each of these attributes should be considered during the initial feasibility study.

The ability to leverage legacy systems is important as both a cost saver as well as a compatibility issue. Users are accustomed to the level of service and features that they currently have, and any new system should build on these and offer more, not less.

Bandwidth control and service provisioning allow higher profitability for the capital investment. Gopalakrishnan points out, "Service providers that make optimal use of their resources are more profitable. Bandwidth control and rapid service provisioning are among the important functions that network equipment must now perform."[1]

It is never too early in the design process to consider the software and necessary hardware for traffic engineering. "As optical networks begin to deliver almost unlimited bandwidth to end users, service providers need to deliver differentiated services dynamically."[1] That is, the providers need to be able to reconfigure their services and bandwidth between nodes on an hour-by-hour or even minute-by-minute basis.

The hardware and software of a network design must support whatever accounting system is used to estimate costs and bill customers. This may not be trivial, as new ways to charge users are evolving. Obviously, the accounting must be accurate and quick.

Unfortunately, new networks must be compatible with other company's networks and those of legacy systems. This can be particularly irritating and limiting for the designer, as even the old legacy hardware (copper, coax, and old fiber) must be supported to allow almost universal coverage.

Reference

1. S. Gopalakrishnan, "Building 10 Gigabit Ethernet/DWDM Metro Area Networks," *Proceedings of the National Fiber Optic Engineers Conference*, pp. 835–836, 2001.

PACKET DELAY TIME

$$t_d = t_s + \left[\left(\frac{p}{1-p} \right) \left(\frac{t_f}{1-G} \right) (1+G) \right] + \frac{t_f}{2} \left(\frac{1+G}{1-G} \right) + 0.00003 \text{ in seconds}$$

where t_d = average packet delay in seconds
  t_s = equivalent t_f for the last cell in the packet

  $p = \dfrac{P_\# t_f}{1-G}$ where $P_\#$ = number of packets arriving per second

  t_f = transmission frame time
  G = geometric distribution parameter (usually between 0.1 and 0.75)

Discussion

As each ATM cell is sent down a labyrinthine path though a network, delays can occur at switches and routers. The total delay a cell experiences is simply the addition of the time of each individual delay.

The above is based on a *fiber to the home (FTTH)* system, assuming a Poisson distribution with a rate of $P_\#$ of packets arriving per second, and that the total delay time is the simple addition of all of the contributing delays. As Salloum points out,

We realize that some of these terms (or contributing delays) are independent of one another. This is due to the fact that they happen in different transmissions frames. This is an important fact to realize because it simplifies the calculations of Laplace transforms, noting that the Laplace transform of many statistically independent functions equals the products of the transforms of each function.[1]

This allows one to derive the simplified equation above.
' The 30 µs added in the back are to account for the 10 µs for positioning the packet in the transmission frame (typical for fiber optics to the home) and 20 µs for a propagation delay.

The above equation is quite sensitive to the geometric distribution parameter, G. As G increases, the delay also increases as each packet contains more ATM cells. G is a function of the number of users connected.

Reference

1. H. Salloum, "Performance of ATM over Multipoint Access Systems," *Proceedings of the National Fiber Optic Engineers Conference,* 1996.

COMPARISONS OF KAUTZ HYPERNET PERFORMANCE INDICES

For undirected hypernets,

$$\gamma_{avg} = \frac{r\bar{H}}{\Delta}$$

where γ_{avg} = average hyperedge load
r = size of the hyperedge
Δ = degree of the hypernet (or hypergraph)
$\bar{H}$ = average logical hop count

For directed hypernets,

$$\gamma_{avg} = \frac{s\bar{H}}{d}$$

where s = a positive integer representing the outsize
d = a positive integer representing the "out-degree" of the net

And for the both directed and undirected case,

$$\mu_{avg} = \bar{H} - 1$$

where μ_{avg} = average load processed per node

Discussion

The above are the relationships for some theoretical performances figures of merit for network load and flow when analyzed using Kautz techniques.

The above is based on Kautz hypernets and fundamental flow conservation. The derivation assumes hypernets with uniform carried traffic and $1/(N-1)$ units of traffic flowing between each source-destination node pair. It is further assumed that each source injects one unit of traffic into the network, regardless of the network size, and that the throughput is N.

H is an important performance parameter, as it determines the throughput, the time delay, and the processing load. And Stern and Bala point out that "the performance results are 10 to 15 percent better when the shortest-path-on-edge routing is used."[1]

Reference

1. T. Stern and K. Bala, *Multiwavelength Optical Networks*, Prentice Hall, Upper Saddle River, NJ, pp. 570–586, 2000.

SHANNON'S THEOREM

For a digital communications network, the maximum rate of transmission cannot exceed

$$C = BW \log_2(1 + SNR)$$

where C = maximum data rate in bits per second
  BW = electrical bandwidth
  SNR = signal-to-noise ratio

Discussion

Claude Shannon (sometimes called "the Richard Feynman of the information sciences") was perhaps the first person to realize, analyze, and publish the fact that communications can all be digital. From that insight, he ignited the science and engineering of digital information. Shannon, who died in Feb. 2001, developed many fundamental ideas of information theory in the 1940s and published the above theory in 1948.[1]

Shannon's theorem bounds the upper limit for the amount of information that a measurement contains. The above rule implies that power is inversely correlated with the exponential of bit rate. The expansion of bandwidth geometrically increases communications.

As stated by Bairstow,

> In a brilliant analysis, Shannon showed that any communication channel had a speed limit, measured in bits per second. Above that theoretical limit, perfect transmission was not possible. Below that limit, Shannon demonstrated that perfect transmission would be possible no matter how much distortion and static there might be in that channel...[2]

The above is valid assuming that a sufficiently sophisticated error correction code is employed.

Generally, a high-signal-to-noise data stream can carry more information. As Hobbs points out, a frequency measurement of one cycle at a given SNR (or CNR) will yield b bits of accuracy where b is[3]

$$b = (1/2)(\log_2(SNR)) + \log \pi$$

As Gilder states,

> Digital communications efficiency declines as power increases.... Increased electrical power means more dispersion and nonlinearity in fiber and more interference in the air. Reducing power expended per bit enables exponentially increasing bit rates. The rise of digital

implies a constant preoccupation with reducing the power usage of every component of the information infrastructure.[4]

This, of course, ignores the ever-present considerations of noise.

Bairstow tells an interesting information story about Shannon and the mechanical information paradox that he built.

Once he built a lidded box with a single switch on the outside. When the switch was flipped, a mechanical hand rose from under the lid and turned the switch off. The hand then retreated into the box and the lid closed, leaving the gadget in the same state as before.[2]

Thanks for the metaphor, Claude.

Similar devices were actually sold in the 1960s as a child's toy; many had a flywheel inside so that they could complete their task and fully retract after the switch was turned off.

References

1. C. Shannon, "A Mathematical Theory of Communications," *Bell System Technical Journal,* July 1948.
2. J. Bairstow, "The Father of the Information Age," *Laser Focus World,* Feb. 2002, p. 114.
3. P. Hobbs, *Building Electro-Optical Systems, Making It All Work,* John Wiley & Sons, New York, p. 614, 2000.
4. G. Gilder, *Telecosm,* The Free Press, New York, p. 266, 2000.
5. J. Westland, "The Growing Importance of Intangibles: Valuation of Knowledge Assets," *BusinessWeek CFO Forum,* 2001.

SNR AFTER N OPTICAL AMPLIFIERS

The signal-to-noise ratio after N optical spans is approximately

$$\text{SNR} \cong 58 + P_{out} - L - NF - 10 \log_{10} N \text{ in dB}$$

where SNR = optical signal-to-noise ratio in decibels after N spans, measured in a 0.1-nm bandwidth

P_{out} = output power per channel in dBm (remember, a dBm is a decibel referenced to 1 mW of laser power, and input coupling loss should be included in the NF)

L = span loss in dB

NF = noise figure in dB

N = number of spans

Discussion

The above relationship allows for a quick calculation of the receiver's optical SNR (in dB) for a signal traveling through a number of optically amplified lengths of fiber.

The receiver's "signal-to-noise ratio can be increased decibel for decibel by increasing the output power per channel, by decreasing the noise figure, or by decreasing the span loss."[1] The SNR falls off proportionally to the log of the number of spans; thus, the preferred way to increase SNR is to decrease span loss for multiple-span systems.

Reference

1. J. Zyskind, J. Nagel, and H. Kidof, "Erbium-Doped Fiber Amplifiers for Optical Communications," in *Optical Fiber Telecommunications*, I. Kaminov and T. Koch, Eds., Academic Press, San Diego, CA, p. 37, 1997.

GUIDELINES FOR NETWORK ARCHITECTURE TOPOLOGY

When building a network, it is advisable to plan the topology of the management enterprise so that the distance across the network between any two far end nodes is constant.

Discussion

The above rule results in balancing out the response times and balances the distance (and therefore the timing of one packet to the next packet, regardless of the path that they take).

Sharman further states,

> This balances out the response time across the network and guarantees fairness with respect to distance or number of hops a packet must cross to get from one side of the network to the other. The farther the receiver is from the sender, the more hops the packet must cross. In building routed networks, the best topologies turn out to be the ones having an architecture with the highest bandwidth at the core, a limited number of hops, and a consistent diameter across the end-to-end topology.[1]

In addition, Sharman also gives the following general guidelines for a SONET data communications network (DCN):

Guideline 1.   All SONET node elements (NEs) in the same configuration with connected data communications channels (DCC) (e.g., a ring network) should reside in the same OSI routing area.

Guideline 2.   A gateway network element (GNE) and at least one central office (CO) router should reside in the same routing area.

Guideline 3.   Data communications switching routers should not share a routing area with a central office (CO) router.

Guideline 4.   Redundancy should be provided through multiple switching routers in separate physical locations with different routing areas, and multiple CO routers and WAN links to every GNE, so no single point of failure exists in the data communications network.

Guideline 5.   SONET node elements deployed without a direct WAN interface should be accessible from a management system via more than one gateway network element (i.e., primary and secondary GNEs should be provided) to protect against a single gateway network element point of failure.

Guideline 6.   As a starting point, OSI routing areas should be limited to 50 routing devices (all SONET node elements and all data communications routers) until experience validates the actual numbers. (*Note:* some vendors' equipment may support a greater number of devices within an OSI routing area.)

Reference

1. C. Sharman, "Guidelines for Sonet DCN Architecture Engineering," *Proceedings of the National Fiber Optic Engineers Conference,* 1999.

USEFUL TRAFFIC RELATIONSHIPS

$$G = S(1 + E)$$
$$G = Mp_s$$
$$S = p_s G$$
$$E = (1 - p_s)/p_s$$
$$p_s = (1 - p)^{M-1} \approx e^{-G} \text{ for } M \gg 1$$
$$D \approx (1 + E)t$$

where G = average packets per channel-slot (the total normalized traffic offered to the network)
  M = number of independent queues
  S = average packets per channel (normalized throughput)
  E = average number of retransmissions of a packet until successful
  p_s = probability that a transmission is successful
  p = probability that a station transmitted into a slot
  D = average delay until successful reception
  t = source to destination propagation time (usually the dominant quantity for optical networks)

Discussion

The above simple relationships allow one to quickly calculate throughput and its effect on the network, given the probability that a transmission is successful, based on some rational assumptions for networks.

The above assumes that one needs to keep track of and differentiate between original traffic and retransmitted traffic. The above also assumes a paradigm in which lost packets are retransmitted, which is common for datacom. In some applications, lost packets are simply dropped. Thus, G contains both original packets and regenerated packets. It also assumes that there is no queuing delay.

Unfortunately, in the real world, the larger G becomes, the larger E tends to be, and there is a point at which E increases faster (as the network becomes unstable and begins to fail). Stern and Bala state,

> This illustrates the well known difficulty with all systems involving packet loss: The retransmitted traffic puts an extra load on the network, which may reduce significantly the maximum throughput. To avoid congestion and instabilities, offered traffic must be kept well below the value that maximizes throughput, and special transmission control algorithms must be used to maintain stability.[1]

Reference

1. T. Stern and K. Bala, *Multiwavelength Optical Networks*, Prentice Hall, Upper Saddle River, NJ, pp. 348–350, 2000.

DESIGNING DWDM SYSTEMS

Component selection and installation care is essential to assure that dense wavelength division multiplexing (DWDM) will function as desired.

Discussion

When installing DWDM, it is important to take care of a few key things. These include making sure that all of the components in the system are selected for, and adjusted for, operation at the wavelengths of the specific channels chosen. In addition, transmitters and receivers must operate at exactly the right wavelengths. Careful installation and selection of components should assure that crosstalk between channels is low. Finally, it is important that adverse effects in fibers, like dispersion and nonlinear behavior (including four-wave mixing), are properly managed. A good resource for determining requirements for frequency deviation, frequency drift, frequency wander, jitter, and chirp can be found in *ITU Recommendation G.692*.[2]

We can be even more specific about some of the key performance parameters of DWDM. We list them below.

1. *Optical rejection ratio* (sometimes called dynamic range). It is essential that the system design and the diagnostic tools used with it are able to sense small signals in the presence of large signals in adjacent channels. Strength differences might be as much as 35 dB.

2. *Receiver sensitivity.* Similarly, detectors and diagnostic equipment must be able to detect very low signal levels. Said another way, the system components must add very little noise to the system. Otherwise, full characterization of the system performance may not be possible.

3. *Resolvable bandwidth.* Test equipment must be able to clearly resolve the signals from closely spaced wavelength bands. For modern DWDM, this means being able to resolve 12.5-GHz channel spacings.

4. *Wavelength accuracy.* Finally, we must be assured that all components in the system, as well as test equipment, are calibrated accurately. Without such calibration, the likelihood of channel-to-channel conflict is dramatically increased.

References

1. E. Gagnon, "Field Testing DWDM Systems," *Proceedings of the National Fiber Optic Engineers Conference,* 1997.
2. ITU Recommendation G.692, *Optical Interfaces for Multichannel Systems with Optical Amplifiers,* Defense Information Systems Agency, Washington, DC, http://www-comm.itsi.disa.mil.

WDM AND TDM

Wavelength domain multiplexing (WDM) and time division multiplexing (TDM) are complementary.

Discussion

Dense wavelength division multiplexing (DWDM) increases the bandwidth by adding more laser lines (or colors). TDM increases the bandwidth by adding more bits per second. An improvement in TDM can be placed on every color of a DWDM system. Conversely, an addition of a given laser line can be added to a given TDM system. Thus, they tend to be complementary (although some nonlinear effects can start to cause cross corruption in extremely long links of extremely high bandwidths).

SETTING BIT ERROR RATE

Great care should be taken in setting the threshold of a system. Setting it too low results in poor bit error rate (BER) when system noise is high, and setting it too high demands high signal power.

Discussion

The optimal quantization threshold is a function of transmitter system design and the response of the receiver circuits to duty cycle asymmetry. When designing a network, following the above rule seems to give optimal overall performance. Reduce the transmitting power (which increases reliability and reduces the number of amplifier stations) to the lowest level that gives a noise acceptable to the selected BER to have a system with lower cost, good performance, and robust transmission at lower (but acceptable) noise levels.

Reference

1. D. Fishman and B. Jackson, "Transmitter and Receiver Design for Amplified Lightwave Systems," in *Optical Fiber Telecommunications*, I. Kaminow and T. Koch, Eds., Academic Press, San Diego, CA, p. 91, 1997.

A TERABIT IS A TERRIBLE THING TO WASTE

A terabit per second (Tb/s) is approximately equal to

- 20 to 60 million simultaneous two-way phone calls
- 1 million books per second downloaded
- 300 years of a daily newspaper downloaded per second
- 160,000 streaming NTSC videos
- 800,000 streaming CD quality audio links
- 160 bits per second for every human

Discussion

DWDM networks hold the promise to transmit terabits per second over a single fiber, and tens of terabits per second over a multifiber cable. Consider a 384-wavelength DWDM system pumping data down a fiber at 40 Gb/channel. This is equal to over 1.5 Tb/s. If a cable contains 100 fibers; this is 150 Tb/s. However, switching, receiving, and demultiplexing the 38,400 laser lines are still formidable challenges.

This data rate is required to allow streaming video to many simultaneous users, and it is important to comprehend the amount of data that can be transmitted in such networks.

Not considering overhead and redundancy, telephones use about 7 kHz for each direction—that's 14 kHz per two-way call. Twenty million of those calls gives 280×10^9 b/s, and 60 million is about 0.840 Tb.

Obviously, the authors are being glib about wasting a terabit. As pointed out by many, the ability to waste bandwidth is a hallmark of the state of technology that a society has achieved. We all hope for a day when a wasted terabit is of no concern, much as a wasted megabyte of computer memory is of no consequence today. Twenty-five years ago, virtually no computer had a megabyte of memory in total.

A petabit is a thousand times a terabit....

USE FIBERS RATHER THAN COPPER FOR ENTERPRISE DISTRIBUTION

Avoid using copper between buildings; use fibers instead.

Discussion

Everyone knows about the potential for lightning damage to above-ground communications cables. One can also consider wireless technology, but the costs are much higher, and reliability can be an issue. This is especially germane for enterprise networks. Also, let us remember it is the twenty-second century, and copper and fiber are not the only solutions.

Network Elements

Most optical telecom systems rely on fibers. But they rely on much more. The fibers themselves convey the light, but its creation, switching, detection, and management require a number of other components, such as those described in this chapter.

Among the topics discussed here are the components that act as the "glue" in these systems. For example, we discuss switches, their performance as a function of bandwidth, and their isolation performance. Similarly, we offer some insight into the operation of Benes and Clos switch architectures and their operation.

The devices covered in this chapter include common network components such as high-speed detectors (or receivers), high-speed erbium-doped fiber amplifiers (EDFAs), and switches. Although fast, modulated laser transmitters qualify as network components, we found enough related rules to justify creating a separate chapter dealing with them.

Not all of the rules are about specific elements; rather the rules define how performance features of those elements interact to define system performance. For example, we have included several rules that relate to the fundamental performance limits of switches and other components, including bandwidth, rise time, switching speed, channel-to-channel isolation, and noise performance. A particularly useful rule provides a quick way to determine the relative importance of shot noise and Johnson noise.

EDFAs receive considerable attention in this chapter. The authors recognize that many of the rules that are included here might have also been placed in the optical fiber chapter but, for a number of reasons, we have found it useful to collect these specific rules here. Mostly, this was done because the material captured in the rule defines how EDFAs interact with other system elements to form systems and, although both a laser and fiber, the EDFA functions as an amplifier, generally in a rack along with other network components.

Similarly, detectors are addressed here in the system context. For example, one of the rules defines a general goal for the signal-to-noise ratio that is desired when one uses avalanche photodiodes in a communication system. This rule derives from experience in using these detectors in real systems, rather than studies of the detectors in an isolated laboratory evaluation. Indeed, it is our goal that these rules provide insight into how system components should be deployed to ensure system success.

To continue to emphasize the system nature of the rules in this chapter, we have included information on the expected reliability of particular components.

In a sense, this chapter describes a set of components that are likely to be overcome by emerging technology on a scale of five to ten years. In nearly every case, the topics addressed here (with the exception of the basic ideas of bandwidth) relate to components that are being developed or that are being replaced by new inventions. A review of the recent history of EDFAs, detectors, switches, and similar items shows that most of the installed technology is two to three generations old and is being rapidly displaced by new inventions. Since the pace of technology development has been accelerating, ever more rapid replacement can be expected. The biggest trend is to integrate multiple network elements into a single device, such as transmitter lasers and receiving detectors, performance monitors and EDFAs, and switches and add/drop subsystems.

An example of a network element that has evolved tremendously over the past 30 years is the transmitter. Early short-range systems could exploit multimode fibers and LED devices. With the advent of laser transmitters and a new generation of single-mode fibers, the early system concepts quickly gave way. Today's systems go even farther by exploiting the high gain and low noise performance of EDFAs and a new generation of detectors integrated with a vertical cavity surface emitting laser (VCSEL) on a chip.

In summary, the user of this book will have to be diligent to monitor the emergence of new technologies that may replace the ones discussed in this chapter.

ONE THOUSAND PHOTONS

A good avalanche photodiode has a sensitivity of about 1000 photons per bit.

Discussion

The equation below shows how to make this computation.

$$\overline{N}_P = \frac{B}{\eta R_b} Q^2 \left[F + \frac{1}{Q} \sqrt{\frac{8\pi k T C_e}{q^2 M^2}} \right]$$

where N_P = number of photons per bit
  B = electrical bandwidth
  R_b = bit rate; B/R_b is usually about 0.7 (note that the best that can be achieved is 0.5, meaning that the bandwidth of the receiver is one-half the bit rate)
  C_e = effective capacitance of the detector (around 0.1 pF) and is equal to $1/2\pi(R_b B)$
  Q = signal-to-noise ratio (usually $\approx$ 6) and the BER is $\approx 10^{-9}$
  M = gain, usually $\approx$ 10
  η = quantum efficiency of the receiver (around 0.7)
  q = electron charge
  F = excess noise factor for the amplification process and ranges from 3 to 10 for most detector materials

Note that the expression is more accurate for a PIN diode (when M and $F = 1$), since the equation derives from a Gaussian approximation for the thermal and photon noise terms, even though the latter is a Poisson process. Excess noise in avalanche photodiodes (APDs) is not Gaussian, so the approximation is better for PIN diodes, where excess noise is not present.

Reference

1. I. Jacobs, "Optical Fiber Communication Technology and System Overview," Chap. 2 in *Fiber Optics Handbook*, M. Bass, Ed. in Chief, and E. Van Stryland, Assoc. Ed., McGraw-Hill, New York, p. 2.9, 2001.

ADD/DROP 25 PERCENT

It is desirable to be able to drop and add up to about 25 percent of the channels on a fiber at each add/drop node.

Discussion

The applicability of DWDM has been accompanied by a need for the ability to add or drop signals at intermediate points. To have a system in which the demands of customers can be accommodated, it needs to have complete flexibility in the configuration of add/drop and pass-through. The flexibility to perform laser line adds and drops is necessary for efficient network operation. In the middle of the night, there is lower voice and data demand, and thus fewer lasers are needed per fiber than during the day. Thus, throughout the day, laser lines are added and dropped dynamically, depending on the demand. Additionally, if using an optical switch, two lasers of the same wavelength cannot be sent down the same fiber; this may result in one being "dropped." The above-stated 25 percent rule sets a typical standard for the amount of lasers that might be added or dropped.

Thus, if you are planning a 100-line DWDM system, the rule indicates that you should design in the ability to add and drop about 25 of those lines at each node.

Reference

1. W. Tomlinson, "Comparison of Approaches and Technologies for Wavelength Add/Drop Network Elements," *Proceedings of the National Fiber Optic Engineers Conference*, 1998.

AVALANCHE PHOTODIODE PERFORMANCE

The multiplication factor for an avalanche photodiode (APD) can be estimated from

$$M^{-1} = C\left(1 - \left(\frac{V_b}{V_{bd}}\right)^m\right)$$

where M = multiplication (gain) factor
 C = a material (sometimes device dependent) constant
 V_b = bias voltage
 V_{bd} = breakdown voltage
 m = an empirical constant (values between 1.4 and 6 have been observed)

Discussion

An APD can generate many electron-hole pairs from a single interaction with a photon. It happens this way: when a photon interacts, it generates a single electron-hole pair. These objects travel in opposite directions through the lattice under the influence of the bias voltage and can produce other electron-hole pairs (an amplification effect) by collision with other electrons and holes. The multiplication (or gain) factor M is the average number of electron-hole pairs produced from a single initiating electron-hole pair.

Gain can be very high (~100) for silicon APDs, in which breakdown voltage is about 400. Other materials provide less gain, but M equal to 10 to 20 can be generally expected.

References

1. T. Limperis and J. Mudgar, "Detectors," Chap. 11 in *The Infrared Handbook*, Wolfe, W., and Zissis, G., Eds., ERIM, Ann Arbor, MI, pp. 11-36 and 11-37, 1978.
2. M-K. Ieong, "The P-N Junction," EE3106 Lecture 7, p. 14, available at http://www.ieong.net/ee3106/Lecture7.PDF.
3. E. Garmire, "Sources, Modulators, and Detectors for Fiber-Optic Communication Systems," Chap. 4 in *Fiber Optics Handbook*, M. Bass, Ed. in Chief, and E. Van Stryland, Assoc. Ed., McGraw-Hill, New York, p. 4.76, 2001.

AVOIDING PUMP LOSS

An isolator (preferably one that is polarization insensitive) placed at the 32 percent of the whole length of the active fiber can partially compensate for the DWDM insertion loss.

Discussion

In dense wavelength division multiplexing (DWDM) systems that employ erbium-doped fiber amplifiers (EDFAs), insertion loss can be a problem. The erbium-doped fiber amplifier has been a tremendous resource for fiber system designers. The happy coincidence of the gain properties of such amplifiers with the high transmission window in modern fibers has enabled long-distance systems with low noise and high bandwidth.

One issue with such amplifiers, however, is the back propagation of amplified spontaneous emissions (ASEs). In extreme cases, this signal can saturate components near the input point of the fiber. That is, the small input signal from a laser source is overwhelmed by the back-propagating and amplified ASE, which has traveled through the entire gain section of the EDFA. In practice, ASE causes degradation of the noise properties of the system.

For some time, there has been interest in distributing the gain sections of EDFA with isolation sections between them so that back propagation does not accumulate and does not reach the input end of the fiber. Reference 2 shows experimental results wherein the noise figure of the fiber system is at a minimum when the position of the isolator ranges from about 20 to 60 percent of the length of the fiber. The authors conclude that, in general, isolators should be placed 1/3 to 1/2 the length of the fiber from the input end and that the position should be adjusted as a function of isolator loss. As loss increases, the isolator should move away from the input end.

It can also improve the noise and gain performances of the amplifier—even more so if input and output isolators are used to protect the amplifier from external back-reflections.

References

1. F. Pozzi, D. Moro, and M. Maglio, "2.5 Gbit/s Repeaterless Fiber Optic Link: Optimization and Testing of a Remotely Pumped Erbium-Doped Fiber Amplifier Realized with Commercial Optical Components," *Proceedings of the National Fiber Optic Engineers Conference*, 1996.
2. S. Yamashita and T. Okoshi, "Performance Improvement and Optimization of Fiber Amplifier with a Midway Isolator," *IEEE Photonics Technology Letters*, 4(11), p. 1276, Nov. 1992.

COMPONENT RELIABILITY

Obviously, the reliability of a systems depend on the reliability of its individual components. The table summarizes some example performance data.[1]

FIT Rates of the Internal Components of an Optical Fiber Amplifier (OFA)

OFA components	Type	FIT range
Pump laser module	E/O, FO, E	1000–5000
Active fiber module	OF	100–1000
Filter	FO	300–500
PD	E/O	300
PBC	FO	100
Splitter	FO	50
Electronics	E	100–500
Isolator	FO	300

Note: A FIT is one failure per 10^9 operating hours.

Discussion

This rule describes the reliability of components that fall into three categories commonly used in optical telecom systems: fiber optics (FO), electro-optics (E/O), and electronics (E). In presenting the data, we use the term FIT, which is one failure per 10^9 operating hours. We also refer to a polarization beam combiners (PBCs) and photodiodes (PDs).

These results derive from field experience and data obtained from manufacturers. These numbers are impressive, but one needs to remember that millions of these devices are deployed, leading to system-wide failures every 100,000 hours or so, which is about every 11 years.

A similar set of data are provided in Ref. 2 as follows.:

Network element or device	Failure rate (FIT)*	Unavailability†	MTTR (hr)
Optical fiber amplifier	2,485	$U_{OFA} = 1.49 \times 10^{-5}$	6
Fiber cable‡	114/km–570/km	U_{FC}	variable
1:2 coupler	50	$U_C = 10^{-7}$	2
Transmitter	180	$U_T = 3.6 \times 10^{-7}$	2
Receiver	70	$U_R = 1.4 \times 10^{-7}$	2
Network node	5,000	$U_{node} = 10^{-5}$	2

*1 FIT = 1 failure per 10^9 hr.
†$U = $ FIT $\times$ MTTR/10^9 hr.
‡This corresponds to 1 to 5 cable breaks per year in a 1,000-km network.

References

1. T. Del Giorno, S. Etemad, and D. Pipan, "Reliability of Optically Amplified WDM Transmission Links," *Proceedings of the National Fiber Optic Engineers Conference*, 1997.
2. L. Wosinska and L. Pedersen, "Scalability Limitations of Optical Networks Due to Reliability Constraints," *Proceedings of the National Fiber Optic Engineers Conference*, 2001.

ELECTRICAL FREQUENCY BANDPASS FOR A SPECTROMETER

The electrical frequency bandpass for a spectrometer should be

$$f \geq 0.8RS$$

where f = electrical frequency bandpass
R = resolving power of the spectrometer, given by the wavelength divided by the incremental wavelength bandpass ($\lambda/\Delta\lambda$)
S = scan rate in scans per second

Discussion

Spectrometers are frequently employed in DWDM systems as the heart of optical power monitors, test equipment, and other network elements. The electrical frequency bandpass must be $0.8RS$ so that the detection circuit can keep up with the changes in detector output; the $\geq$ symbol is included to indicate that enough bandwidth is needed to catch nuances in the spectrum.

This rule assumes the integration time is at least five times the system time constant. In addition, this rule is true only if the resolving power is defined as the number of spectral lines per scan.

The sensor response is a exponential function related to the rise time. When the integration time is equal to 2.2 times the rise time, the hardware output achieves 80 percent of the theoretical (square wave) amplitude of the source. When the integration time is 5 times the dwell or integration time, the output achieves 99.3 percent of the amplitude and represents a nominal design value for accurate systems.

The electrical frequency response is related to the time constant by a simple RC network. The integration time is a function of the resolving power and the scan rate. All this can be combined to yield a frequency response equal to $5RS/2$, which can be simplified to the above.

In general, design practices dictate that the time constant be equal to or less than the dwell time divided by 5. This results in a loss of instrument sensitivity, but the time constant can be increased after the measurement by postprocessing; however, it cannot be decreased without a signal-to-noise penalty.

Reference

1. C. Wyatt, *Radiometric Calibration: Theory and Methods,* Academic Press, Orlando, FL, pp. 145–148, 1978.

Kartalopoulos's Advantages and Disadvantages of EDFAs

Kartalopoulos gives the following paraphrased advantages of erbium-doped fiber amplifiers (EDFAs):

1. High power transfer efficiency from pump to signal (>50 percent)
2. Directly and simultaneously amplify a wide wavelength region (in the region of 1550 nm) at an output power as high as + 37 dBm, with a relatively flat gain (>20 dB)
3. Saturation output greater than 1 mW (10–25 dBm)
4. Long gain time constant (>100 ms) to overcome patterning effects and intermodulation distortions
5. Large dynamic ranges
6. Low noise figures
7. Transparent to optical modulation format
8. Suitable for long-haul applications
9. Newer EDFAs able to operate in the L-band

Additionally, Kartalopoulos's has the following paraphrased disadvantages of EDFAs:

1. EDFAs are not small devices (the associated fibers are kilometers long) and cannot be easily integrated with other semiconductor devices.
2. EDFAs exhibit amplified spontaneous emission (ASE) of light, which has an adverse effect on performance of the system.
3. Crosstalk exists.
4. Gain saturation exists.

Discussion

Erbium-doped fiber amplifiers are simply fibers that provides optical gain. The erbium is pumped by another laser (generally at 980 or 1480 nm) to raise the erbium atoms into an excited state. As the data signal propagates through the laser, it triggers the erbium atoms to release photons of the same wavelength, resulting in an amplification of the signal light. To obtain several decibels of gain, this phenomenon generally must occur over several meters of fiber. Generally, the fiber is not part of a cable but, rather, part of a set of rack-mounted equipment in a telco hotel or a device at a splice in a cable. The EDFA "box" consists of inputs for the signal fibers, a pump laser with several (maybe 20 or 30) meters of fiber wrapped in a spool, and the output.

EDFAs have matured from a laboratory discovery into a commercially viable and mature product in less than 12 years—a much shorter time than the typical technological cycle for electro-optical products. AT&T and the University of Southampton first reported the phenomenon in 1987. Oki Electric produced the first EDFAs for telecommunications, and by 1999 they were in widespread use in ultra-long-haul and long-haul networks. Over $3.3 billion in EDFA sales were reported in 2000, and the market is expected to exceed $7 billion within the next five years or so.

Reference

1. S. Kartalopoulos, *Introduction to DWDM Technology*, IEEE Press, Piscataway, NJ, p. 124, 2000.

NUMBER OF ELEMENTS FOR BENES SWITCHES

$$\text{No. switches} \cong n\log_2 n - \frac{n}{2}$$

where n = the number of permutations required (or input to output possibilities)

Discussion

Generally, a Benes switch can be made from the above number of 2×2 switches; thus, this is based on 2×2 switches. Optical switches tend to be costly, electronic switches consume a lot of power and are inefficient, and both require space in the telco hotel. It is always advantageous to create the required number of switch possibilities (or permutations) with the fewest actual hardware switches.

The Benes architecture is one of the most efficient switch designs with respect to the number of permutations of crosspoints. The simplest switch architecture is a simple square array of switches in which a series of vertical lines intersect a series of horizontal lines, and each intersection can be open or closed. For this crossbar architecture, the number of switches grows rapidly as n^2, which is more switches than necessary to create all the permutations. The Clos architecture adds a middle step between the inputs and outputs for a multistep architecture and, generally, a nonblocking Clos switch can be made with approximately $5.6n^{3/2}$ crosspoints (Stern and Bala, p. 42).[1] The Clos architecture can be factored recursively to produce a series of 2×2 switches and produce a Benes fabric as shown in Fig. 6.1. Note that the figure provides 36 permutations with only 9 switches.

Stern and Bala[1] also point out (p. 43) that electrical switches are either closed or open, while optical switches tend to leak a little (say, 5 percent).

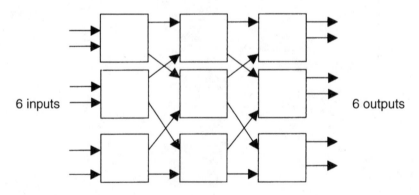

6 inputs

6 outputs

Figure 7.1 Clos architecture.

Thus, optical switches require some special architectures with extra binary elements to reduce crosstalk.

References

1. T. Stern and K. Bala, *Multiwavelength Optical Networks.* Prentice Hall, Upper Saddle River, NJ, pp. 40–43, 2000.
2. Lecture, G. Sasaki, Univ., of Hawaii, p. 13, http://www-ee.eng.hawaii.edu/~sasaki/EE693F/Spring01/Lectures/lectseto.pdf.

SWITCHING TIMES

The length of the fibers can limit switching time in large networks that include switching to avoid faults.

Discussion

Figure 6.2 illustrates how the switching time varies with fiber length and the number of optical cross connects (OXCs) on the network. The curves assume the following parameters:

- Defect detection and alarm indication signal (AIS) time = 125 ms
- AIS relay time at each "passthrough" OXC = 125 ms
- AIS detection time at the end node OXC = 375 ms
- Path protection switch time at the selector of the end node OXC = 20

Reference

1. M. El-Torky and F. Kamoun, "Designing Large Scale Optical Mesh Networks," *Proceedings of the National Fiber Optic Engineers Conference*, 2001.

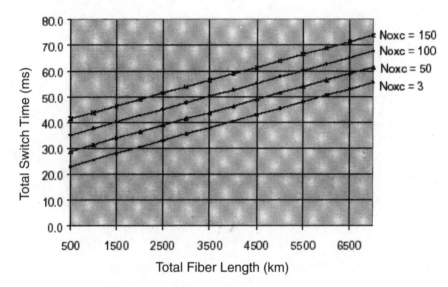

Figure 7.2   Switch time vs. fiber length.

PHOTONS-TO-WATTS CONVERSION

To convert a radiometric signal from watts to photons per second, multiply the number of watts by the wavelength (in microns) and by 5×10^{18}.

$$\text{Photons per second} = \lambda(\text{microns})(\text{no. of watts})(5 \times 10^{18})$$

Discussion

Actual results are only for a given wavelength or an infinitesimally small bandpass. Typically, using the center wavelength is accurate enough for lasers or bandwidths of less than 0.05 µm. However, if doing this conversion for wide bandpasses, set up a spreadsheet and do it in increments (say, 1/100 µm).

This rule is valid only if the wavelength is expressed in microns. The constant must be adjusted for wavelengths expressed in other units. If you require more than two significant figures, the constant is 5.0345×10^{18}.

The foundation of this rule is the amount of energy contained in a photon. This is given is the common expression, $E = h\nu$. Since watts are ergs/s, we can compute the actual conversion as

$$\text{Watts} = \frac{hc}{\lambda} \text{ photons/s}$$

Therefore,

$$\text{Photons per second} = \frac{\text{watts} \times \lambda}{hc} = \text{watts} \times \lambda \text{ (in meters)} \times (5 \times 10^{24})$$

for all terms with the dimension of meters (wavelength in meters). There are a million microns in a meter, so the constant is 10^6 smaller if one uses microns.

RECEIVER SPEED

The more sensitivity exhibited by a receiver, the smaller its bandwidth.

Discussion

As is true in almost all electronics, one cannot get something for nothing. This rule derives from the well known effects of expanding the bandwidth of any type of detector—not just the ones used in telecom. The foundational physics involves the conversion of photons (in this application, the ones coming through the fiber and associated network elements) into electrons at the detector. The quantum efficiency of the process, while high, is not perfect, leading to statistical noise in the output electrical signal. This process is defined by Poisson statistics, in which the variance in the signal is equal to the mean of the signal. This means that the highest SNR that can be achieved is equal to $\sqrt{N}$, where N is the rate of creation of electrons. Said another way,

$$N = I\Delta t/e$$

where I is the current created by the received light. In interval Δt, this current creates a charge, $I\,\Delta t$ (measured in coulombs). When divided by the charge of a single electron, e, we get the number of electrons produced in the process. The square root of the above expression is the signal-to-noise ratio (SNR). If we also note that the bandwidth of the receiver process is $\Delta f = 1/2\Delta t$, we find that SNR can be expressed as

$$\text{SNR} = \sqrt{I/2e\Delta f}$$

Another way to write SNR is to divide the mean signal (in this case, I) by the standard deviation of the noise. It is evident from the expression above that this means that the standard deviation of the noise is $\sqrt{I2e\Delta f}$. That is, the larger the bandwidth, the larger the noise current and the lower the signal-to-noise ratio. Said another way, it takes a lower bandwidth to control the noise current and provide a low-noise/highly sensitive system. Hence, the rule.

INCREASE POWER FOR RELATIVE INTENSITY NOISE

Relative intensity noise (RIN) in diode sources requires extra power to be put into the system.

Discussion

Relative intensity noise results when stray light is reflected into the diode sources in the system. A good approximation for the additional power required to overcome this effect is captured in the following equation:

$$P_{RIN} = -5\log\left[1 - Q^2(BW)(1 + M_r)^{2g}10^{RIN/10}\left(\frac{1}{M_r}\right)^2\right]$$

where Q = system signal-to-noise ratio
  BW = receiver bandwidth
  M_r = receiver modulation index
  g = a number between 0 and 1 that relates the *RIN* and the optical power in the system

RIN is measured in dB/Hz. When managed properly, *RIN* can be less than 0.5 dB.

Reference

1. C. DeCusatis and G. Li, "Fiber-Optic Communication Links (Telecom, Datacom, and Analog)," Chap. 6 in *Fiber Optics Handbook*, M. Bass, Ed. in Chief, and E. Van Stryland, Assoc. Ed., McGraw-Hill, New York, p. 6.15, 2001.

RISE TIME

The switching time characteristics of nonamplified junction photodetectors such as PIN detectors is determined by the RC performance down to rise times of 10 ns. In practice, the rise time is approximately

$$T_r = 2.2RC$$

where T_r = rise time in seconds
R = resistance in ohms
C = capacitance in farads

Discussion

In optical telecom, the rise time of a detector used as a receiver is of primary importance. As network data rates move to 40 Gb (and faster) per second, the detector's rise time must be able to respond to this pace. As can be seen from the above equation, to be useful at these fast data rates, the detector should have very low effective capacitance and resistance. In addition, the amplifier used in these systems must have short rise times as well. Low-input-impedance amplifiers usually have high noise. To manage this situation, transimpedance amplifiers are employed with high feedback resistance. This creates a situation in which the detector diode capacitance determines the bandwidth of the system. Much work is ongoing to manufacture low-cost, low-capacitance InGaAs detectors. This is based on electrical engineering of photodetectors and any other circuit that has capacitance and resistance.

The above equation assumes a rise time between 10 and 90 percent of peak response.

$$T_r = T(90\% \text{ of response}) - T(10\% \text{ of response})$$

$$0.1 = 1 - e^{-t/RC}$$

$$0.9 = 1 - e^{-t/RC}$$

$$e^{t/RC} = 1/(1 - 0.1) \text{ and } e^{t/RC} = 1/(1 - 0.9)$$

$$e^{(t-t)/RC} = (1 - 0.1)/(1 - 0.9)$$

$$T_r = T_1 - T_2 = RC\ln[(1 - 0.1)/(1 - 0.9)]$$

$$T_r = 2.197 \ RC$$

This derivation may not be valid for extremely short rise times, as other effects may contribute the shape of the pulse. In most cases, this rule can be used to compute features of the network when the rise time is measured.

This rule is useful for estimating a detector's rise (or response) time for optical communication or scanning systems. Often, these systems result in very short effective integration times, so rise time can be critical. It is also useful for estimating the performance of photodetectors in optoelectronic applications.

The response time of a photodetector should be much less than the dwell or integration time. This assures that the full collection of light signal is accumulated and is not suppressed by the period during which the circuit is responding.

Lloyd[2] points out that many photodetectors behave like a single RC low-pass filter in this respect, with the first equation dominating.

References

1. A. Chappell, *Optoelectronics Theory and Practice*, McGraw-Hill, New York, p. 187, 1978.
2. J. Lloyd, *Thermal Imaging Systems*, Plenum Press, New York, p. 108, 1979, and information supplied by Dr. Robert Martin, 1995.

SHOT NOISE RULE

If the photocurrent from a photodiode is sufficient to drop 50 mV across the load resistor at room temperature, the shot noise equals the Johnson noise.

Discussion

This rule can be proven by investigating the comparison of Johnson noise (caused by the random motion of carriers within a detector, usually thermal in nature) and shot noise (the result of the statistics of photon-to-electron conversion process in the detector). The rms Johnson noise voltage is expressed as

$$\sqrt{4kTR\Delta f}$$

where k = Boltzmann's constant
T = temperature of the resistor
Δf = bandwidth of the detector

The noise bandwidth is generally $1/2t_i$, where t_i is the integration time.

The shot noise (as is proven elsewhere in this chapter) is the product of the noise current and the resistance defined above. That is, the shot noise voltage is $R\sqrt{I2e\Delta f}$. Here, I is the average current in the detector, and e is the charge of an electron.

If we now take the ratio of the Johnson noise to the shot noise, we find

$$\sqrt{\frac{2kT}{RIe}}$$

Now we can compute this ratio for $T \sim 300$ K, and the voltage drop $RI \sim 50$ mV. The value of Boltzmann's constant is presented in the appendix. Care must be taken with the units. The easiest way to work through the problem is to use cgs units in which volts have the units of $cm^{1/2} gm^{1/2}$ per second, charge has the units of $cm^{3/2} gm^{1/2}$ per second, and energy (joules) has the units of cm^2 gm per second squared. Using these values, we find the ratio to be nearly unity.

This rule is derived from many sources. Two examples are listed in the references.

References

1. P. Hobbs, *Building Electro-Optical Systems, Making It All Work,* John Wiley & Sons, New York, p. 117, 2000.
2. T. Limperis, "Detectors," in W. Wolfe and G. Zissis, Eds., *The Infrared Handbook,* ERIM, Ann Arbor, MI, 1978.

STERN AND BALA'S SWITCHING RULES

1. The cost of a wavelength-selective switch typically increases at least linearly with the number of segments of the spectrum it switches independently.
2. It is often desirable to route a bundle of wavelength channels on a common path through the network.
3. Network management at the level of optical nodes is simplified if there are a small number of wavebands to keep track of instead of a large number of wavelength channels.

Discussion

These DWDM-hostile rules are based on the state of the art, which makes it difficult to route and switch multiple wavelengths. Imagine a fiber cable coming into a telecom hotel with 64 fibers and 384 colors on each fiber. The result would be a large number of wavelength channels to route and switch. If the fiber leaving the telecom hotel has the same capability, then there are "x" connections. That is a lot of fibers to deal with, and a lot of connections to track to be certain that you don't switch two identical wavelengths down the same send fiber.

Because of rule 1, Stern and Bala[1] point out, "Regardless of the wavelength density on the fiber, it is less costly to switch/route the optical signals as a small number of aggregated groups of channels, with each group contained in a contiguous waveband, rather than as a large number of individual λ-channels."

The second rule is especially valid for multipoint optical connections. The third rule is "analogous to the management advantages in traditional telephone networks, accruing from the use of digital cross-connect systems for provisioning a high-bandwidth digital infrastructure."[1]

Reference

1. T. Stern and K. Bala, *Multiwavelength Optical Networks*, Prentice Hall, Upper Saddle River, NJ, p. 30, 2000.

TEMPERATURE SENSITIVITY OF AN ERBIUM-DOPED FIBER AMPLIFIER

The temperature sensitivity of an erbium-doped fiber amplifier (EDFA) is approximately –0.023 dB/°C in the temperature range of 25 to 95°C.

Discussion

The gain of an EDFA often falls as the temperature increases. As DWDM increases the density of electronic equipment in telco hotels, the operating temperature also rises. Also, many EDFAs are buried in the ground, where there is little airflow for cooling.

Temperature sensitivity is generally due to the misalignments of the fiber couplers that are inherent to the design, but slight effects from absorption and changes in the indices of refractions may also contribute.

An EDFA's gain as a function of temperature can be expressed as

$$\Delta G = P_{pi}\left(\frac{dG}{dP_p}\right)\Delta\eta_p + G\Delta\eta_s + P_{si}\left(\frac{dG}{dP_s}\right)\Delta\eta_s$$

And $\Delta\eta$ is the change of coupling efficiency per °C, approximated by

$$\Delta\eta \approx -4\eta_o\left(\frac{d_o\Delta d}{w_1^2 + w_2^2} + \frac{(n_c^2\theta_o\Delta\theta + n_c\theta_o^2\Delta n_c)(2\pi w_1 w_2)^2}{\lambda^2(w_1^2 + w_2^2)}\right)$$

where ΔG = temperature sensitivity of the gain, change per °C
P_{pi} = input pump power
(dG/dP_p) = gain slope with respect to the pump power
(dG/dP_s) = gain slope with respect t the signal power
P_{si} = signal power
$\Delta\eta_p$ = power coupling loss (assume on the order of –0.001 per °C)
G = EDFA gain
$\Delta\eta_s$ = signal coupling loss (if unknown, assume approx. –0.0005 per °C).
$\Delta\eta$ = change in coupling efficiency per °C (due to thermal expansion)
η_o = total power coupling efficiency (including radial and angular misalignments)
d_o = initial radial misalignment
Δd = change in radial misalignment
w_1 = beam width of the first fiber in the coupler
w_2 = beam width for the second fiber in the coupler

n_c = refractive index of the core
θ_o = initial angular misalignment
$\Delta\theta$ = increment of the angular misalignment (if unknown, assume about 5.4×10^{-5} rad/°C)
λ = free-space wavelength

Liu et al. indicate[1] that the last term of the ΔG equation can be neglected under many conditions and that the slope of (dG/dP_p) is usually temperature dependent, but it can become mostly temperature insensitive if you choose the pump power and fiber length carefully.

The drop for is almost linear from 30 to about 60°C, then the curve becomes highly nonlinear.

Reference

1. C. Liu et al. "Temperature and Electromagnetic Effects on Erbium-Doped Fiber Amplifier Systems," *Optical Engineering,* 37(7), pp. 2095–2100, July 1998.

RAMAN GAIN

The gain from a Raman amplifier increases almost linearly with the wavelength offset between signal and pump, peaking at about an 100-nm difference, then it drops off rapidly.

Discussion

In 1928, Sir Chandrasekhara Venkata Raman experimentally observed that, when light is scattered, photons are created at slightly different wavelengths (both higher and lower). A molecule absorbing a photon, and then re-emitting a photon while transferring some rotational or vibration energy (to or from) the photon, causes the effect. The additional wavelengths are characteristic of the material doing the scattering and thus result in a powerful spectroscopic tool.

This phenomenon can be employed with transmission fibers to create an amplifier within the transmission medium and is typically referred to as a *distributed Raman amplifier (DRA)*. A pump beam is co-launched into a fiber with a wavelength slightly lower than the signal to be amplified. Amplification occurs when the pump photon yields its energy to create a new photon at the signal wavelength, with the energy difference going to vibrational energy. This amplification tends to have a function defined by the above rule.

TEN-DEGREE DROP

Every 10°C drop in temperature will reduce silicon PIN photodiode noise by 30 percent.

Discussion

The noise of a typical photoconductor or photovoltaic bandgap detector can be reduced by reducing its temperature. This is because there are statistically fewer electrons with the thermal energy needed to cross the bandgap at a colder temperature, and thus fewer noise electrons that might be mistaken for signal electrons generated by a photon. From room temperature, for a typical silicon PIN photodiode, this results in a decrease in noise of 30 percent for every 10°C drop for the first few tens of degrees; beyond that, diminishing returns are to be expected.

Reference

1. P. Webb, from Weiss, S., "Rules of Thumb," *Photonics Spectra*, Oct. 1998, pp. 136–139.

POLARIZATION OF ERBIUM-DOPED FIBER AMPLIFIERS

The gain of an erbium-doped fiber amplifier (EDFA) is polarization independent.

Discussion

An EDFA amplifies the signal as a function of pump wavelength, not polarization. It will equally amplify an S or P polarization and is agnostic as to a combination.

980 VS. 1480

Typically, 980 nm pumping results in a noise figure 1 dB lower than that for 1480 nm pumping.

Discussion

EDFAs require optical pumping, and pumping near the modulation bands is the most efficient, so often pumping occurs at 1480 nm. However, typical EDFAs can also be pumped (and frequently are) at 980 nm. The shorter wavelength results in less noise.

ERBIUM-DOPED FIBER AMPLIFIER OUTPUT POWER

The output power of an erbium-doped fiber amplifier (EDFA) is proportional to the pump power.

Discussion

Erbium-doped laser amplifiers allow amplification of the modulated laser light carrying the data. An EDFA has an output power approximately proportional to the input power when signal levels are high and the amplifier is saturated. When the amplifier is saturated, pump absorption from the ground state is balanced by stimulated emission from the first excited state induced by the signal. This can be verified by examining the detailed rate equations of the populations of the erbium energy levels. The result is an approximately proportional relationship between the pump power and output power.

References

1. E. Desurvive et al., "Efficient Erbium-Doped Fiber Amplifiers at 1.55 μm Wavelength with a High Output Saturation Power," *Optics Letters,* 14(22), p. 1266, 1989.
2. C. Bradford, "Optical Amplifiers Come of Age," *WDM Solutions,* Feb. 2001.
3. A. Willner and Y. Xie, "Wavelength Domain Multiplexed (WDM) Fiber-Optic Communications Networks," *Handbook of Optics,* Vol. 4, M. Bass, Ed., McGraw-Hill, New York, pp. 13.17–13.23, 2001.

RAMAN FIBER GAIN

Single-mode fibers have a Raman gain of about 10^{-13} m/W.

Discussion

This indicates that Raman gain can occur with normal fibers and should be considered for long links when exact measurements are required. The physical effect results from interaction between the guided wave and high-frequency phonons in the material. To put this in perspective, if the link consisted of about 6.5 trips to the sun, the signal would experience a gain to 1 W. Conversely, a link from the Earth to the moon and back would result in an output of just under 1 mW.

Reference

1. T. Brown, "Optical Fibers and Fiber-Optic Communication," Chap. 1 in *Fiber Optics Handbook,* M. Bass, Ed. in Chief, and E. Van Stryland, Assoc. Ed., McGraw-Hill, New York, p. 1.38, 2001.

AMPLIFIER SPONTANEOUS EMISSION NOISE POWER

Amplifier spontaneous emission (ASE) noise power depends on amplifier gain as described by the following equation:

$$P_{ASE} = n_{sp}\frac{hc}{\lambda}B_o(G-1)$$

where G = amplifier gain
B_o = optical bandwidth
n_{sp} = spontaneous emission factor (1 for ideal amplifiers)
h = Planck's constant
c = speed of light

Reference

1. I. Jacobs, "Optical Fiber Communication Technology and System Overview," Chap. 2 in *Fiber Optics Handbook*, M. Bass, Ed. in Chief, and E. Van Stryland, Assoc. Ed., McGraw-Hill, New York, p. 2.12, 2001.

NOISE BANDWIDTH OF DETECTORS

When unknown, the noise bandwidth of a photodetector is commonly assumed to equal 1 divided by 2 times the integration time, or

$$N_b = 1/(2t_i)$$

where N_b = noise bandwidth
t_i = integration or dwell time

Discussion

This rule is based on simple electrical engineering of photodetectors and simplification of circuit design. This rule is highly dependent on detector material, architecture, and optimal filtering of noise. It assumes a rectangular pulse and a value for the noise cutoff of 3 dB down from the peak.

The rule does not always properly include all readout and preprocessing signal conditioning effects, and can vary from $1/t_i$ to $1/4t_i$. The noise bandwidth of a detector is a function of signal processing, semiconductor physics, and readout architecture. The response of the electronics (amplifiers and filters) to this will have a more gentle rise related to the inverse of the bandwidth. For a well designed system, the electronics can be matched to provide a minimal noise bandwidth as approximated above.

RESPONSIVITY AND QUANTUM EFFICIENCY

An optical detector's responsivity (in amps/watt) is equal to its quantum efficiency divided by 1.24 times the wavelength in microns.

Discussion

This occurs because of the definition of quantum efficiency. It is defined as the number of electrons generated per second per incident photon on the active area of the detector at a particular wavelength. When making the conversion (which invokes Planck's constant, the number of electrons per Coulomb, and the speed of light), the number 1.24 appears if wavelength is expressed in microns.

Chapter

8

Noise

In this chapter, we introduce a number of rules related to the influence of noise in the performance of optical communication systems. In this discussion, we must make a distinction between the definition of signal-to-noise ratio (SNR) as commonly used is the electrical engineering context and the analogous definition for optical systems.

In electrical circuits, especially ones that deal with communications (as opposed to conveying power), the designer must evaluate the system to know if the signals are going to be detectable. The most common way to do this is to estimate the signal level and some statistically derived measure of the noise. The SNR is then simply the ratio of two numbers: the mean signal power and the standard deviation of the noise power. For completeness, we mention that noise is expected to have a zero mean value, a Gaussian distribution, and a randomness that allows the total variance in a system to be computed by adding the variance of each of the noise terms. Remember that variance is the square of the standard deviation. Keep in mind that this definition applies to electrical signals as might be found in wires, amplifiers, and other electrical circuits. As a reminder, the SNR in electrical systems is defined by the power of the signal and the statistics of the noise.

Now, consider the case of interest in optical communications; we need to be confident that sufficient light representing the signal is present, compared with noise present in the fiber in the form of unwanted light (such as light from four-wave mixing, ASE, spurious laser emission, etc.). Later, we will need to take note that the signal levels in a communication system are temporally binary—either a "1" or a "0" after the receiver (including detector noise and pulse dispersion) in the time domain. With this in mind, we define the optical signal-to-noise ratio (OSNR) as independent of time domain and receiver effects; the former are inherent to the "eye diagram." Thus, for DWDM systems, both an OSNR and "eye" are required for proper specification and troubleshooting, as they are basically independent.

First, lets review how light is electronically measured. Photons of light each carry a small amount of energy, found by multiplying their optical frequency by Planck's constant. The rate of arrival of photons at a detector defines the power in that stream, since power (watts) is the rate of change of energy. In almost all OSNR test-equipment and receivers, no direct measurement of optical power takes place; the optical signal is converted to electrical signals in a detector. OSNR detectors are similar to those of the receiver, except they operate much more slowly and with lower noise. The rate of arrival of photons at the OSNR detector defines the optical power and is converted to electrical power by the detector. The measure of performance of the detector is its responsivity, which has the units of amps/watt. In photovoltaics and photoconductors, commonly used in optical telecom, watts of power are delivered by the photons and are converted to electrical current in amps or a potential difference in volts.

The OSNR is measured in the optical domain. This measurement compares the peak power at a given wavelength point to the power a given wavelength distance from the peak (as shown later in Fig. 8.6). Often, the latter measurement point is the minimum between the peak and its next-nearest information-carrying neighbor in wavelength. Sometimes, it is the midway (in wavelength space) between the two DWDM data channels. Some instruments allow it to be adjustable, allowing one to make a plot of OSNR as a function of distance from an ITU defined wavelength.

Now that we have theoretically defined OSNR, we must discuss the measurement "points" of the previous paragraph. Obviously, an infinitesimally small point in wavelength would mean a zero optical bandwidth and result in no power on the detector. The instrument measuring the OSNR must have a bandwidth of a finite width. Generally, the test equipment will scale its measurement to be equivalent to a measurement taken with a 0.1-nm bandwidth. Modern instruments have very sensitive detectors with narrower bandwidths, and older systems have wider bandwidths—both normalized to 0.1 nm. Thus, for a DWDM application, one might measure different OSNRs depending on the instrumentation. Other considerations being equal, the one with the narrowest bandwidth is the most accurate.

Now we can discuss the eye-diagram's "Q." We have already pointed out that modern communication systems use binary coding; only two states can exist. Each has its own signal and noise levels. That is, there are actually two distinct SNRs, one for a "1" and one for a "0." Q is a parameter that captures information about both SNRs in the time domain. It does so by making a measure of the difference between a "1" and a "0" and comparing that difference to the noise in the system. The plot of this is called an "eye diagram" (as discussed below). Incidentally, the "eye diagram" does more, as it also shows the rise and fall (in the time domain) of the "1" and "0." In equation form, Q looks like

$$Q = \frac{y_1 - y_0}{\sigma_0 + \sigma_1}$$

where y_1 = signal when a 1 is being sent
   y_0 = signal when a 0 is being sent
   σ_0 = standard deviation of the 0 signal
   σ_1 = standard deviation of the 1 signal

We can see from this formulation that, if the signal levels for a "1" and a "0" are close, or if each has a lot of noise, Q will be small. Large Q assures (either because the signal levels are very distinct or because the noise level is low) that a "1" and a "0" will not be confused. The high-Q case is desirable, since it assures that the intended signal level is correctly received. As we will see below, and in subsequent rules in this chapter, Q can be used to make estimates of bit error rate (BER).

One key indicator of signal quality and of noise in an optical telecom system is the *eye pattern*. This is a plot of the measured signal (generally in amperage, but it could be volts or photons or other units) of a bit meant to represent a one overlaid with a bit meant to represent a zero. These frequently take the form of images from an oscilloscope, optical channel analyzer, or some other practical electronic measurement test equipment. When well defined, the shape of the curve resembles a human eye; hence the name *eye diagram*, as shown in Figure 8.1a. In Figure 8.1a, the zero portion is low and well defined, with little noise on the signal, while the one is high, rises steeply, and again has little noise. When such signals are sent down a fiber and through amplifiers, switches, and other network elements, they will be corrupted. The corruption can take many deleterious forms, including having the variance of the highs (for one) and lows (for zero) increased, the distance between the ones and zeros can be decreased, and the slopes may become more gradual. Figures 8.1b

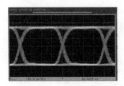   

a.   b.   c.

Figure 8.1   (a) Clean and (b, c) corrupted eye diagrams (from Ref. 1).

and 8.1c show highly corrupted eye diagrams from self-phase modulation and dispersion. An eye diagram essentially illustrates Q in a graphic way.

One of the wonderful attributes of digital communication is that relatively noisy signals can be read accurately. The receiving electronics must only determine if the signal was meant to be a one or a zero. This determination can be made on rather poor-looking eye diagrams, with only an occasional error. The rate at which errors occur is referred to as the *bit error rate (BER)*, which is the frequency of mistaking a one for a zero or vice versa. In a typical system, it is desirable to have a BER lower than 10^{-9}. Employing techniques such as forward error correction and calibration of the expected signal can greatly reduce the bit error rate for a given SNR. The reader will find several rules in this chapter describing interesting shortcuts to estimating BER and its effects.

In this chapter, we include rules relating to amplitude noise. Another chapter deals with other types of phenomena that degrades the optical signal. These include phase noise, timing jitter, polarization effects, and a number of others. The rules in this chapter are the result of a large body of work available in many papers and books.

Reference

1. A. Mitchell, Z. Zhu, and X. Zhang, "Advances in Repeaterless Systems," *Proceedings of the National Fiber Optic Engineers Conference*, 1999.

HAVE AT LEAST 10 DB OF MARGIN

Good design practice requires that there be at least 10 dB of margin between the receiver sensitivity and the optical signal, or

$$\frac{P_R}{NEP} \geq 10 \text{ dB}$$

where P_R = power at the receiver
  NEP = noise equivalent power of the receiver

Discussion

The optical signal at the receiver should be greater than that required to achieve the necessary bit error rate (BER). This is to compensate for inaccurate vendor specifications, minor errors, overly optimistic loss values, and forgotten factors, and to allow some margin for future maintenance. The loss margin is important, as it defines the amount by which the received optical power exceeds the required amount. Reference 1 suggests that it should always be 10 dB or more. Others are comfortable with a lower value, e.g., 3 to 6 dB.

In the real world, the estimated values for each component are always too optimistic or uncertain by some amount. These amounts add up and result in an underestimation of what is required at the receiving end. This 10-dB margin has proved useful and reliable in recent applications.

Table 8.1 gives some typical values for the signal loss in a link. If one assumes that an LED or laser source has 3 dB of power, then the entries in the table can be subtracted from it, resulting in –24.25 dBm of power at the receiver. If the receiver sensitivity is –40 dBm, then the system has 15.75 dB of margin.

TABLE 8.1 Typical Values of Signal Loss in a Link

Link	Signal reduction	Comment
Source coupling loss	5 dB	Due to reflections, area mismatch, etc.
Transmitter-to-cable connector loss	1 dB	Transmitter to fiber optic cable with ST connector, the value accounts for misalignment
Splice loss	0.25 dB	
Fiber optic cable attenuation	20 dB	
Cable-to-receiver connector loss	1 dB	Assumes an ST connector and includes misalignments

In addition, Ref. 2 and another rule provide similar advice. Signals arriving at a receiver should be between −10 and −5 dBm (0.1 to 0.3 mW) for each fiber channel and have an SNR of 100. Keep in mind that these numbers are dependent on the BER requirement, optical filtering, electrical filtering, and many other factors.

References

1. A. Girard et al., *Guide to WDM Technology and Testing,* EXFO Electro-Optical Engineering Inc., Quebec City, Canada, pp. 97–98, 2000.
2. D. McCarthy, "Growing by Design," *Photonics Spectra,* July 2001, p. 88.

PROBABILITY ERROR AND BER AS AN EXPONENTIAL FUNCTION OF Q

The bit error rate (BER) is an exponential function of Q and can be approximated by

$$P \approx \frac{1}{Q\sqrt{2\pi}} e^{-(Q^2/2)}$$

where P = probability of error
Q = noise factor of an eye diagram, which is equal to the difference in signal (e.g., amperage) or eye opening between a 1 and 0 signal, divided by the sum of the noises of the signals

or

$$Q = \frac{y_1 - y_0}{\sigma_0 + \sigma_1}$$

where y_1 = signal when a 1 is being sent
y_0 = signal when a 0 is being sent
σ_0 = standard deviation of the 0 signal
σ_1 = standard deviation of the 1 signal

Discussion

Q is defined as the eye opening divided by the noise of the signal. In an eye diagram such as the one in Figure 8.2, the total opening, $y_1 - y_0$ can be considered to be the signal amplitude, while the variance in that signal is the instantaneous noise (not including crossover from the next or pre-

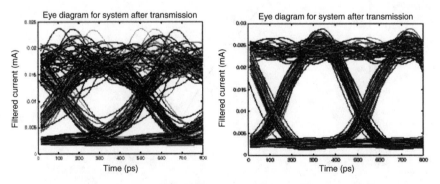

Figure 8.2   Eye diagrams. The left-hand image is much noisier (with a Q of ≈ 3) than the right-hand one (with a Q of ≈ 10) (from Ref. 6).

vious bit interval) or variation in the lower signal level (the 0) and the upper signal level (the 1). Often, Q is converted and referred to in terms of decibels.

The above assumes that the noise is Gaussian. A Gaussian assumption is a good approximation, but only an approximation, as sometimes the noise may be dominated by a failing component exhibiting a non-Gaussian behavior. Nevertheless, a Gaussian assumption for the noise distribution results in the use of the complementary error function (erfc) to estimate the probability of error, and such is commonly used to derive the above approximation from

$$P = \frac{1}{2}\text{erfc}\left(\frac{Q}{\sqrt{2}}\right)$$

Thus, it turns out that the BER and SNR can be calculated and related in the same manner as in target detection for radar, sonar, and imaging systems. Stern and Bala point out that setting of the decision threshold that minimizes the probability of error (or BER) is at the crossover of the two bit intervals.[1] This corresponds to a maximum likelihood (ML) detection. So when the "eye" has a Q (signal-to-noise factor) of 5, the BER is about 10^{-6}, and when it has an SNR of 5, it is between 10^{-4} and 10^{-5}; a Q of 6 is required for a BER of 10^{-9}. The shape of the curve is nearly exponential for all but the noisiest cases as shown in Figure 8.3. Generally, one needs a

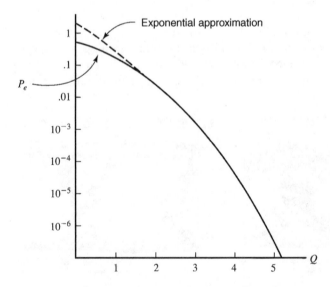

Figure 8.3 BER as a function of Q shows nearly exponential dependence (from Ref. 1).

BER in the neighborhood of one part per thousand for digital voice applications, and 10^{-6} to 10^{-13} for data transmission.

The above assumes no forward error correction, as this can make the behavior non-Gaussian and improve the error rate (see associated rule in this chapter).

References

1. T. Stern and K. Bala, *Multiwavelength Optical Networks*, Prentice Hall, Upper Saddle River, NJ, pp. 215–218 (Fig. 2 on p. 218), 2000.
2. G. Bosco and R. Gaudino, "Toward New Semi-Analytical Techniques for BER Estimation in Optical System Simulation," *Proceedings of the National Fiber Optic Engineers Conference*, 2000.
3. S. Sridharan, M. Jarchi, and A. Schmitt, "A 2.5 GBit/s Reed-Solomon FEC System for DWDM Optical Networks," *Proceedings of the National Fiber Optic Engineers Conference*, 2000.
4. C. DeCusatis and G. Li, "Fiber Optic Communication Links (Telecom, Datacom and Analog)," in *Handbook of Optics*, Vol. 4, M. Bass, Ed. in Chief, McGraw-Hill, New York, pp. 6.2–6.7, 2001.
5. J. Pecar, R. O'Connor, and D. Garbin, *Telecommunications Fact Book*, McGraw-Hill, New York, p. 23, 1993.
6. A. Michell, Z. Zhu, and X. Zhang, "Advances in Repeaterless Systems," *Proceedings of the National Fiber Optic Engineers Conference*, 1999.

CARRIER-TO-NOISE RATIO AS A FUNCTION OF MODULATION INDEX

The carrier-to-noise ratio can be expressed as

$$CNR \cong \sqrt{\frac{\pi}{2}} \mu^{-3} \exp\left(\frac{1}{2\mu^2}\right)$$

where CNR = carrier-to-noise ratio
  μ = root mean square (rms) modulation index

Discussion

The carrier-to-noise ratio is a legacy figure of merit resulting from cable TV and voice telecom, but it is sometimes used in optical telecom. It is defined as the square of the signal current divided by the variance of the noise, generally the clipping noise power at the receiver. At low levels, it can be dominated by noise caused by thermal fluctuations; at high noise levels, it is generally limited by the shot noise of the receiver or the laser transmitter, or fiber amplifier noise.

Stern and Bala[1] state that the above "assumes that the modulation signal consists of M sinusoidal carriers of unit amplitude. For the case of a large number of M and random phases, it can be approximated as a Gaussian random process," which results in the above equation. The root mean square modulation index can be written in terms of the number of sinusoidal carriers of unit bandwidth and modulation indices as

$$\mu = m\sqrt{\frac{M}{2}}$$

where μ = rms modulation index
  M = number of sinusoidal carriers of unit amplitude
  m = modulation index

Reference 3 gives the following handy equation for relating the OSNR and CNR:

$$OSNR = 2\sqrt{\frac{CNR}{m^2} \frac{B_{sa}}{\Delta \upsilon}}$$

where $OSNR$ = optical signal-to-noise ratio
  B_{sa} = resolution of the spectrum analyzer used
  m = modulation depth of the subcarrier
  $\Delta \upsilon$ = optical bandwidth

References

1. T. Stern and K. Bala, *Multiwavelength Optical Networks,* Prentice Hall, Upper Saddle River, NJ, p. 295, 2000.
2. A. Saleh, "Fundamental Limit on Number of Channels in Subcarrier-Multiplexed Lightwave CATV System," *Electronics Letters,* 25(12), pp. 776–777, 1989.
3. G. Rossi and D. Blumenthal, "Optical SNR Monitoring in Optical Networks Using Subcarrier Multiplexing," *Proceedings of the 26th European Conference on Optical Communications,* pp. 227–228, 2000.
4. I. Jacobs, "Optical Fiber Communication Technology and System Overview," in *Handbook of Optics,* Vol. 4, M. Bass, Ed. in Chief, McGraw-Hill, New York, pp. 2.14–2.16, 2001.

ERBIUM-DOPED FIBER AMPLIFIER SPONTANEOUS EMISSION NOISE FIGURE

$$NF = \frac{SNR_{in}}{SNR_{out}} = \frac{2P}{hv(\Delta v)(G-1)}$$

where NF = noise figure, defined as the SNR at the output divided by the corresponding shot noise of the signal at the input

P = amplified spontaneous emission power in one polarization in bandwidth

h = photon energy

Δv = optical bandwidth

G = amplifier gain

Discussion

The erbium-doped fiber amplifier (EDFA) is a gift from the technology gods. It provides excellent, quiet amplification in just the right wavebands to complement the performance of high-performance fibers. They aren't perfect, however. Erbium laser amplifiers amplify the spontaneous emission wavelengths as well as the modulated wavelength. In addition, Rayleigh scattering is an issue. The amplified simultaneous emission (ASE) signal adds to the background and can degrade the amplified signal, and the above NF is a measure of the degradation of the signal by amplified noise. The source of the noise is the transmitter laser and the fiber amplifier itself. The noise is often the dominant noise source at the receiver, as the detector noise and other noise sources are much lower.

Rayleigh scattering, like ASE, causes light back-propagation in the fiber. Both sources can be amplified during their travels, limiting the available gain for the real signal. Most system designers try several methods to limit the ASE problem, including the use of several amplifier stages rather than a single, large one. Nonlinear effects such as four-wave mixing can also occur.

Three to 4 dB of noise figure is about the current technology limit, with 6.5 dB being more common. To estimate SNR_{out}, the rms noise level is often about 1/5 of the bright band noise amplitude as measured using an oscilloscope.[1]

References

1. W. Johnstone, *Erbium Doped Fibre Amplifiers*, student manual, OptoSci Ltd., Glasgow, Scotland, 2001.

2. J. Zyskind, J. Nagel, and H. Kidof, "Erbium-Doped Fiber Amplifiers for Optical Communications," in *Optical Fiber Telecommunications*, I. Kaminov and T. Koch, Eds., Academic Press, San Diego, CA, pp. 18–19, 1997.

3. J. Buck, "Optical Fiber Amplifiers," Chap. 5 in *Fiber Optics Handbook*, M. Bass, Ed. in Chief, and E. Van Stryland, Assoc. Ed., McGraw-Hill, New York, p. 5.6, 2001.

EMPIRICAL EQUATIONS FOR ESTIMATING BER FROM OSNR

$$(10^9)\, e^{-3.45(OSNR)} \qquad\qquad \text{Eq. (1)}$$

$$(10^{24})\, e^{-4.6(OSNR)} \qquad\qquad \text{Eq. (2)}$$

$$(10^8)\, e^{-2.3(OSNR)} \qquad\qquad \text{Eq. (3)}$$

$$(10^{13})\, e^{-3.9(OSNR)} \qquad\qquad \text{Eq. (4)}$$

where   BER = bit error rate
OSNR = optical signal-to-noise ratio in dB

Discussion

The above four equations should be used only with extreme caution. The relationship between BER and signal-to-noise ratio is elusive, and it really cannot be easily placed in a simple way. If you need to make a quick estimate or sanity check on some claim, these can provide such. Each of the equations was curve-fitted by the authors to data in the references. Note that plugging in the same OSNR gives a range of several orders of magnitude from one equation to another. The above does not include the gains from forward error correction.

The form of the relationship does seem to be exponential, which is not surprising. Refer to an earlier rule in this chapter that nicely explains how an exponential approximation works.

Equations (1) and (2) are from Ref. 1. Equation (1) was for a link with no optical amplification and therefore no ASE. Equation (2) was based on the accumulation of errors from four optical line amplifiers. The difference in OSNR between the two is approximately 4 dB for the same BER. The measurements were made on an eight-channel analyzer (200 GHz spacing) carrying 2.5 Gb/s. Care was taken to eliminate signal distortions caused by optical fibers, and all amplifiers had a constant output power. Equation (1) was derived using OSNR values of 8 to 14 dB, while Eq. (2) used values of 14 to 18.

Equations (3) and (4) are derived from data in Ref. 2. These were for two different SONET vendors and are valid from about 11 to 16 dB of OSNR. The signal was a reference OC-48, and the total power received at the receiver was maintained at a constant –20 dBm. The reference gives data for three vendors. Equation (3) was the best, and Eq. (4) was the worst.

References

1. L. Gillner et al., "Experimental Demonstration of the Ambiguous Relation between Optical Signal to Noise Ratio and Bit Error Rate," *Proceedings of the National Fiber Optic Engineers Conference*, 1998.
2. S. Alexander, "WDM System Considerations," *Proceedings of the National Fiber Optic Engineers Conference*, 1996.

GAIN FROM FORWARD ERROR CORRECTION

Forward error correction (FEC) can provide at least 6 dB of improvement with <10 percent overhead.

Discussion

Forward error correction, which is a form of coding that suppresses errors that arise in communication systems, is becoming more commonplace as data rates reach 40 Gb/s and higher; it is, in fact, becoming a necessary enabling technology. By lowering the effective BER and effectively improving the OSNR, we can realize a significant improvement in span length and channel capacity. The amount of data that needs to be added to accomplish FEC is often referred to as the *overhead* or *tax*. As can be seen in the following table, it is generally in the range of 5 to 10 percent.

FEC coding	FEC overhead addition ratio (%)	Corrected BER for raw BER			
		3×10^{-3}	1×10^{-3}	4×10^{-4}	1×10^{-4}
RS-8 (G.709OTUk)	7.1	5×10^{-3}	1×10^{-5}	5×10^{-9}	1×10^{-14}
BCH-20 (G.709 OTUy)	7.1	1×10^{-4}	3×10^{-11}	1×10^{-18}	$>10^{-30}$
BCH-30	8.5	5×10^{-7}	8×10^{-18}	1×10^{-29}	$>10^{-30}$

Several algorithms are currently provided by vendors. Some of the more common are the Bose-Chaudhuri-Hocquenghem (BCH), Hamming, and Reed-Solomon (RS). The RS code has a reputation of providing good random and burst error correction at a low overhead rate. The RS code comprises n symbols made up of m bits related as $n = 2^m - 1$ and encoded with k information symbols, then decoded on the other side using a polynomial whose roots contain information regarding the location of the errors. RS algorithms also have the property that the higher the code rate r, the lower the overheard of the code: $r = k/n$.

Reference 4 gives the following two relationships of BER for a Reed-Solomon code:

$$BER_{input} = 1 - (1 - P_{se})^{1/m}$$

and

$$BER_{output} = 1 - (1 - P_{ue})^{1/m}$$

where BER_{input} = input BER
 P_{se} = probability of a symbol error
 m = number of bits
BER_{output} = output BER
 P_{ue} = probability of uncorrected errors within a codeword

Figure 8.4 (from Ref. 4) is a theoretical plot of the expected improvement for an RS code with a 6.7 percent overhead. Figure 8.5 (from Ref. 5) is a plot of experimental improvements.

Generally, one can assume about a 5 to 7 dB improvement; however, some experiments have generated much better performance improvements. Bergano et al.[2] assert that with "FEC, a BER of 10^{-5} can be improved to below 10^{-15}, as demonstrated in submarine WDM system experiments," and Vodhanel and Gamelin[1] point out that "even a modest 3 dB increase in effective OSNR can be used to double the number of channels or double the number of spans for a fixed EDFA output." Moreover, Thoen et al.[3] contend that a BER of 8×10^{-5} is reduced to 1×10^{-15} with FEC.

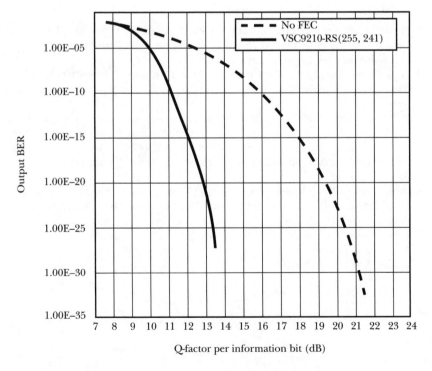

Figure 8.4 Gain for forward error correction (from Ref. 4).

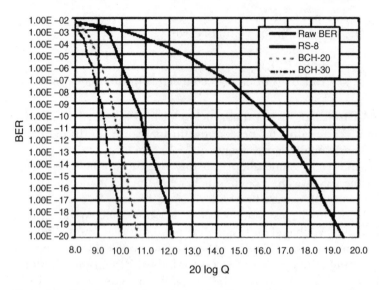

Figure 8.5   Gain for forward error correction (from Ref. 5).

References

1. R. Vodhanel and J. Gamelin, "Large Channel Count WDM Systems: Trade Offs between Number of Channels, Number of Spans and Bit Rate," *Proceedings of the National Fiber Optic Engineers Conference*, 1998.

2. N. Bergano et al., 1996, "100 Gb/s Error Free Transmission over 9100 km Using Twenty 5 Gb/s WDM Channels," *Proceedings of the Conference on Optical Fiber Communications*.

3. E. Thoen et al., "Multi-wavelength 40 Gb/s Transmission Systems for Long-Haul Applications," *Proceedings of the National Fiber Optic Engineers Conference*, p. 529, 2001.

4. S. Sridharan, M. Jarchi, and A. Schmitt, 2000, "A 2.5 Gbit/s Reed-Solomon FEC System for DWDM Optical Networks," *Proceedings of the National Fiber Optic Engineers Conference*.

5. P. Rochon and B. Lavalee, 2001, "Service Transparency and Management of Next Generation 40 Gbps Long-Haul Optical Networks," *Proceedings of the National Fiber Optic Engineers Conference*, pp. 537–548.

6. http://bbcr.uwaterloo.ca/~xjqiu/papers/error/node6.html, 2002.

7. http://grouper.ieee.org/groups/802/3/10G_study/public/july99/azadet_1_0799.pdf, 2002.

NOISE DOMINATION

The noise of an amplified system is dominated by the optical noise produced in the optical generation and amplification, but not by the noise in the receiver.

Discussion

Detectors have become quiet enough that the typical shot noise and thermal noise sources do not dominate the root-sum-square (RSS) system-level resultant noise (see the next rule) of an amplified telecommunications network; thus, they contribute little or nothing to the bit error rate. Currently, the amplification of the modulated light introduces a beat frequency between the modulated signal and that of the spontaneous emission. Both of these phenomena manifest themselves as noise, which are orders of magnitude larger than detector noises.

Since receiver noise is not dominant, this allows the designer to trade the noise specification for other desirable receiver characteristics such as lower inter-symbol interference, improved dispersion tolerance, or lower cost.

References

1. W. Johnstone, *Erbium Doped Fibre Amplifiers*, student manual, OptoSci Ltd., Glasgow, Scotland, 2001.
2. D. Fishman and B. Jackson, "Transmitter and Receiver Design for Amplified Lightwave Systems," in *Optical Fiber Telecommunications*, I. Kaminow and T. Koch, Eds., Academic Press, San Diego, CA, 1997, pp. 89–90.

NOISE ROOT SUM OF SQUARES

Independent noise sources can be added as the root sum of squares. That is, the total variance in the noise of a system is equal to the sum of the variances of the contributing terms, and the standard deviation is the square root of this sum, or,

$$\text{Total Noise Standard Deviation} = \sqrt{n_1^2 + n_2^2 + n_3^2 + \ldots}$$

Discussion

Calculating the combined effect of several error sources in a measurement can be done more easily than most people realize. First, we take note that linear uncertainty of independent noise sources is computed as the root of the sum of their squares, *not* by adding them. Additionally, a rough answer can be had by just taking the largest value. An even better approximation is to multiply the largest value by 1.4. Here is an example that proves the point. Suppose the system has four error terms of 5, 4, 2, and 1. Using the rule above, we would estimate the error as $1.4 \times 5 = 7$. Doing the root sum of squares, we get 6.8.

The random nature of most noise sources encountered in electro-optics is of a Gaussian (normally distributed) form that allows this to be done. This rule is also derived from a common statistical analysis in which the errors in a system are separately analyzed by taking partial derivatives and eliminating the higher-order terms when the noise sources are normally distributed. Any text on error propagation or analysis of experimental data demonstrates this approach.

Many other real "noise" sources are not Gaussian, such as microphonics, noise resulting from 60-cycle electronics, and other periodic sources. Perhaps the most significant "non-Gaussian" noise source is the photon statistics that are important in system noise analysis. Photons are described by Poisson statistics. In cases where Poisson statistics are important, consideration of the fact that the mean and variance are the same must be taken into account. This does not invalidate the equation above. It merely allows the square root to be simplified by simply adding the sum of the mean photon rate of each of the sources. This assumes that the sources of the noise are independent.

Care must be taken in developing an error budget in which all terms are assumed to add as the sum of squares. It is quite common for complex optical systems, particularly those with control systems, to accumulate errors in other than a sum-of-squares way. However, it is almost universally the case that the first analysis that is performed uses this rule to determine the problem areas. A complete accounting of all error sources and the way they accumulate is a complex endeavor, requiring a complete description of the system, which usually doesn't exist until late in the project.

Optical Signal-to-Noise Ratio

The optical signal-to-noise ratio (OSNR) may be calculated from any of the following definitions:

1.
$$OSNR = \frac{P_R}{N_F\{\,G(h\nu)\Delta\nu_o\,\}N}$$

   where P_R = received power
   N_F = noise figure of each amplifier
   G = gain of each amplifier
   $\Delta\nu_o$ = optical bandwidth expressed in frequency (generally measured at 0.1 nm)
   $h\nu$ = photon energy
   N = number of amplifiers

2. The ITU definition of OSNR is

$$OSNR \text{ (in dB)} = P_{out} - L - N_F - 10\log N - 10\log[(h\nu)\Delta\nu_o]$$

   where P_{out} = per channel output power
   L = span loss between amplifiers
   N_F = noise figure of the amplifiers

   and all terms are in dB.

3. Using a less formal definition of OSNR, in terms of P (from Ref. 2),

$$OSNR = \frac{4r(r+1)B_e}{(r-1)^2\nu_o}P^2$$

   where P = square root of the receiver electrical signal-to-noise ratio
   r = signal extinction ratio
   B_e = receiver's electrical bandwidth

4. Reference 2 also gives it, when measured at the receiver, as

$$OSNR = \frac{GP_G}{B_o(G-1)\dfrac{N_F}{2}h\nu\,m_t} = \frac{P_G}{B_o N_F h\nu}$$

   where m_t = number of polarizations, generally 2
   P_G = signal power level
   G = gain $\gg 1$

5. Additionally, Refs. 4 and 5 give the following practical equation based on the effects of the measurement devices:

$$OSNR = \text{worst-case value of } [10 \log (P_i/N_i) + 10 \log(B_m/B_r)] \text{ dB}$$

where P_i = optical signal power in the ith channel
   B_r = reference optical bandwidth, B_m and B_r may be either frequency or wavelength) and is typically 0.1 nm
   B_m = measured bandwidth
   N_i = interpolated value of the noise power measured in the noise equivalent bandwidth such that

$$N_i = \frac{N(\lambda_i - \Delta\lambda) + N(\lambda_i + \Delta\lambda)}{2}$$

where λ_i = measured wavelength
   $\Delta\lambda$ = interpolation offset $\le 1/2$ the channel spacing

Note that there are no multiplications in the numerator; this indicates that the noise is measured at $(\lambda_i - \Delta\lambda)$ and $(\lambda_i + \Delta\lambda)$ and averaged.

6. And Ref. 6 relates the OSNR to the CNR as

$$OSNR = 2\sqrt{\frac{(CNR)B_{esa}}{m^2 \Delta v}}$$

where CNR = carrier-to-noise ratio
   B_{esa} = resolution bandwidth of the electrical spectrum analyzer
   m = modulation depth of the subcarrier

Discussion

As the world moves to increasing optical amplification, switches, and complete optical networks, the measurement and determination of the optical signal to noise is one of the key definitions of optical telecom—and one of the most elusive. The basic methodology for the measurement is illustrated in the Figure 8.6 (from Ref. 4). Ideally, one measures the infinitesimal and exact peak of the communication signal and compares that to the average noise floor between this peak and its neighboring peaks. Much of the confusion results from the ability of an instrument (with a finitely wide optical bandpass) to accurately measure the peak, then to accurately measure the minima between the peaks. Therefore, the OSNR's measured value can actually depend on the measuring equipment characteristics. Other confusions result from which peak (the one of lower wavelength or higher wave-

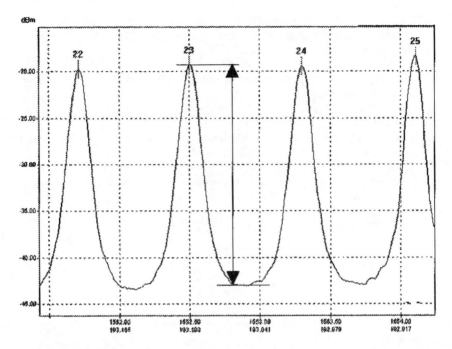

Figure 8.6   The OSNR is a measurement of the maximum and minimum of the optical signal at a defined distance from the peak (from Ref. 4).

length) is to be compared and whether the minimum is based on wavelength or frequency. Also, the minimum may not be between the wavelengths but skewed closer to the longer or shorter peak.

Generally, the measurements are taken or normalized to 0.1-nm spacing. This is obviously incompatible with very dense wavelength packing that is being pursued by several network architects. Fundamentally, as the spacing of the line increases, it will become more difficult to measure the OSNR. This is because it is justly defined as the peak in the bandpass divided by the noise between it and its adjacent DWDM wavelength, as illustrated in the figure. Diffraction and wavelength-pulse broadening effects increase the difficulty of this measurement as the spacing between channels is decreased. This is further aggravated by the desire to have smaller rack-mounted optical monitoring equipment.

The third equation should be used only in a pinch, as the reference indicates. Although it has sound theoretical derivation, it only very crudely meets experimental verification.

References

1. S. Akhtar, "Value of an All-Optical Metro Network Core," *Proceedings of the National Fiber Optic Engineers Conference,* pp. 107–108, 2001.

2. L. Gillner et al., "Experimental Demonstration of the Ambiguous Relation between Optical Signal to Noise Ratio and Bit Error Rate," *Proceedings of the National Fiber Optic Engineers Conference,* 1999.
3. ITU Recommendation G.692, *Optical Interfaces for Multichannel Systems with Optical Amplifiers,* Appendix 1, Defense Information Systems Agency, Washington, DC, http://www-comm.itsi.disa.mil.
4. A. Girard et al., *Guide to WDM Technology and Testing,* EXFO Electro-Optical Engineering Inc., Quebec City, Canada, pp. 99–100, 2000.
5. S. Chabot and G. Levesque, DWDM Impact on Traditional RFTS: Monitoring Colors," *Proceedings of the National Fiber Optic Engineers Conference,* 2000.
6. G. Rossi and D. Blumenthal, "Optical SNR Monitoring in Optical Networks Using Subcarrier Multiplexing," *Proceedings of the 26th European Conference on Optical Communication,* pp. 227–228, 2000.

PROBABILITY OF ERRORS AS A FUNCTION OF BER

The probability that there are no errors in a block of data can be calculated from

$$P_c = (1 - \text{BER})^n$$

where P_c = probability of no errors
    BER = bit error rate
    n = size of the data block in bits

Discussion

The above equation allows one to quickly estimate the probability of an error being in a block of data of n size (e.g., n bits). This is based on classic probability theory, which may be found in any number of texts. Beware of mixing incompatible units for BER and n. The BER can be either the raw BER or the corrected BER, the latter where a forward error correction (FEC) algorithm is used.

Additionally, the probability that exactly one error is generated in a data block of n bits is

$$\text{BER}\,n(1 - \text{BER})^{n-1}$$

and the probability that exactly two errors are found in the block is

$$\frac{n(n-1)}{2}(\text{BER})^2(1 - \text{BER})^{n-2}$$

Furthermore, the reference points out that, for any error correction code, the error-correcting capability depends in part on the block size.[2] That is, for a fixed error-correcting capability and raw BER, the corrected BER will decrease for larger code blocks. Frequently, the code rate increases with larger code blocks. Thus, there is a common and important trade-off between corrected bit error rate CBER and FEC code rate.

References

1. B. Goertzen and P. Mitkas, "Volume Holographic Storage for Large Relational Systems," *Optical Engineering*, 35(7), pp. 1850–1851, 1996.
2. Burle Industries, *The Electro-Optics Handbook*, Burle Industries, Lancaster, PA Inc., pp. 110–112, 1974.

FIBER AMPLIFIER Q-FACTOR

The Q-factor for an erbium-doped fiber amplifier (EDFA) can be crudely estimated from the top plot, and the Q for a Raman from the bottom plot, of Fig. 8.7.

Discussion

In these curves, the pulse energy is shown in femto (10^{-15}) joules, and the pulse width is measured in picoseconds (10^{-12} s).

The Q-factor is dependent on the pulse energy and pulse width of a data stream going through a fiber amplifier. Generally a Raman fiber amplifier will have a higher Q-factor than an EDFA, but EDFAs can be cascaded

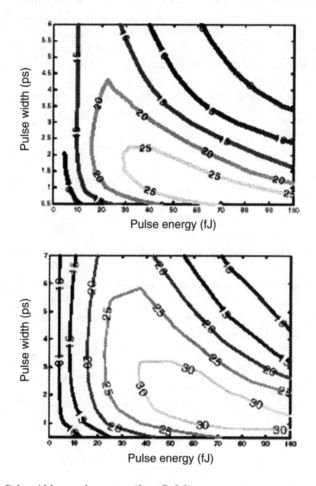

Figure 8.7   Pulse width vs. pulse energy (from Ref. 1).

more confidently. Obviously, the above plots were generated for a specific set of conditions but, in a pinch, they can provide valuable estimates for the hard-pressed practitioner. In general, though, we can expect Raman amplifiers to have a very low noise figure (1–3 dB). They can be pumped anywhere in the single-mode fiber transmission window.

Reference 1 explains the shape of the contours in the following way:

Noise plays an important role for low pulse energies in the left part of the plot. Short pulses have a large spectral width, which decreases the signal-to-noise ratio at the receiver. This, together with the fact that short pulses suffer more from third-order dispersion, limits the performance in the bottom area of the figure. Pulse interactions due to nonlinear effects limit the performance for high energy and long pulses.

The reference also gives the following assumptions used to generate these graphs:

- The data rate is 40 Gb/s.
- They are based on 500 km of fiber with 50 km spans using return-to-zero modulation.
- Pulse energy is defined at the input/output of the fibers.
- EDFAs are assumed to have a noise figure of 4.5 dB.
- The EDFA is placed in front of the dispersion-compensating fiber.
- For the EDFA, the path average pulse energy <E> is used, and <E> is 0.18 E, where E is the pulse energy at the EDFA output.

Reference

1. A. Berntson et al., 2001, "Raman Amplification in WDM Systems," *Proceedings of the National Fiber Optic Engineers Conference*, pp. 840–848.

PEAK POWER IS TWICE

The peak power (P_R) needed to code a bit is *twice* the power contained in the photons that encode that bit.

Discussion

This can be written as the equation below. $\bar{N}_p$ is the average number of photons per bit communicated at wavelength λ, and R_b is the bit rate. The factor of 2 results from the fact that the number of photons per bit is an average, while the power is a peak value, and thus the heart of the rule. Inherent to this equation is an assumed bit error rate (BER) of 10^{-9}.

$$P_R = 2\bar{N}_p R_b \frac{hc}{\lambda}$$

Reference

I. Jacobs, "Optical Fiber Communication Technology and System Overview," Chap. 2 in *Fiber Optics Handbook*, M. Bass, Ed. in Chief, and E. Van Stryland, Assoc. Ed., McGraw-Hill, New York, p. 2.7, 2001.

RECEIVER SNR

Signals arriving at a receiver should be between −10 and −5 dBm (0.1 to 0.3 mW) for each fiber channel and have an SNR of 100.

Discussion

Although this seems on the high side, it allows for a low BER and high confidence of SNR after nominal amplification.

Reference

1. D. McCarthy, "Growing by Design," *Photonics Spectra*, July 2001, p. 88.

9

Optical Fibers

Optical fibers truly represent a photonic revolution that has resulted in the creation of many fortunes. Their enormous transmission capabilities and unique optical properties (e.g., allowing light to be guided around corners) have changed many parts of the electro-optics environment and have been the foremost technical enabler for optical telecom. As commonplace as fiber optics appears to be today, as recently as the 1970s, they were considered exotic. In fact, it was an economic adventure when several firms first offered scientific fiber optic test apparatus in the late 1970s, and the first telecom fiber cable was not lit until 1977.

Fibers are very pure strains of optical material. Generally, telecom fibers are fused silica, typically with one impure molecule in a billion. They are far more transparent than air and as strong as steel. An optical fiber works by playing a quantum electrodynamic (QED) trick: exploiting the total internal reflection and then guiding the light wave along its length—that's why it is called *waveguide*. This total internal reflection results at the interface between two materials of different indices of refraction. An optical fiber is constructed so that the fine strand of transparent material is cylindrically "clad" with a lower-index material. That is, the core has a slightly higher index of refraction than the cladding such that it provides total internal reflection. Usually, an additional surrounding coating and sheavings are provided for protection and strength. Typically, this is referred to as a *buffer* or *protectant*. In practice, fibers can be grouped into coherent *bundles* that can be made to transmit an image, or they can be grouped into incoherent bundles that simply transmit light energy or modulated light energy as in the case of optical telecom.

Often, for telecom fibers, the core is generally doped with GeO_2 or other heavy metal oxides to increase its index of refraction, and the glass core can be doped with other materials for other applications. Telecom fibers typically have an absorption peak near 1383 nm, caused by free OH radicals captured by Si and Ge to form SiO and GeO.

Fibers are frequently classified by the number of modes they transmit. A single-mode fiber is efficient only for a single mode and is frequently used in communications that require long distances and high data rates. This is because they don't experience multimode dispersion. They have diameters less than that of a human hair (about 10 microns), although the cladding is typically about 125 microns.

Multimode fibers allow the transmission of many modes. This type of fiber is usually low in cost, and it is easy to successfully inject (in fiber jargon, this is called *launch*) a pulse into it; however, it suffers from painfully low data rates for long distances. Therefore, these fibers are usually used within telco hotels, within buildings, and for enterprise networks 300 m or less in length.

There are two types of multimode fibers.

- Graded-index multimode fiber, which transmits several modes and is constructed with a radically varying index of refraction
- Step-index fiber, which has a flat index profile (constant) for the core, a constant index for the cladding, and an abrupt interface

In the 1990s, single-mode fiber (SMF) became the dominant fiber for long distance optical telecom, as it eliminated multimode dispersion. In these fibers, fiber manufactures tried to achieve near-zero chromatic dispersion (CD) at 1550 nm with the so-called *dispersion-shifted fiber.* This is important, as even the best sources have a small range of wavelengths. Chromatic dispersion results in these various wavelengths propagating at different speeds, leading to pulse stretching. Some fibers have been optimized to perform well at a narrow band. Unfortunately, optimizing for a single wavelength region tended to have deleterious attributes at wavelengths into which dense wavelength division multiplexing (DWDM) was expanding, so these versions are rarely deployed nowadays. But there exist many kilometers of legacy fiber that was previously deployed. As described by Girard,[1] there are several general flavors of optical fibers, including

- SMF—single-mode dispersion unshifted or nondispersion-shifted fiber (NDSF) (ITU-T: G.652). This has no chromatic dispersion (CD) in the 1310-nm region but a huge chromatic dispersion near 1550 nm and typically a huge attenuation peak in the 1383-nm region. There is SMF with a reduced water peak with 1383-nm attenuation not much higher than the one found in 1310- and 1625-nm regions (ITU-T: G.652C).
- MMF—refers to multimode fiber (ITU-T: G.651).
- DSF—dispersion-shifted fiber (ITU-T: G.653), which has no chromatic distortion in 1550 nm. The problem with G.653 is that when you transmit two or more wavelength channels closely spaced together with typically

closed equivalent peak power and state of polarization, these channels mix to create harmonics. These harmonics (or ghosts) can find their way into the main channels, causing major interferences.

- CSF—cut-off shifted fiber (ITU-T: G.654), which has its cut-off wavelength shifted from 1260 nm to over 1310 nm to ease transmissions in the L-band (1570 to 1610 nm).
- NZDSF—non-zero dispersion-shifted fiber (ITU-T: G.655), which has a little bit of CD (positive or negative) in 1550 nm; some of them have also a reduced CD slope in the 1550-nm region.
- DCF—dispersion-compensating fiber.

The history of fiber optics has taken many interesting turns. There was much research and many beginnings in France in the early nineteenth century. In 1854, John Tyndall demonstrated to the Royal Society that fiber optics hold promise, and in 1880 none other than Alexander Graham Bell proposed telecommunications via light waves. It then took over a century of incremental improvements in the fundamental understanding and manufacturing technology to transform fibers from the realm of a physics laboratory curiosity to an economic juggernaut and household necessity.

The first profitable use of fibers was with clothing and decorations. It is said that Napoleon's funeral drapes were made of optical fibers. Glass fiber dresses and coats were a fashion statement for a while but were apparently too heavy to stand the test of time. Although modern OSHA inspectors wouldn't approve it now, early fibers were frequently drawn by shooting arrows down halls. This technique was perfected by Sir Charles Boys, and he was able to draw surprisingly uniform fibers with less than a 0.2 micron diameter in the late 1880s. He also experimented with using rockets but deemed them "inconvenient."[2]

Crossbows and rockets notwithstanding, these techniques merely produced the core. In the early twentieth century, research concentrated around placing a highly reflective coating (such as silver or aluminum) around the fiber to contain the light. Surprisingly, it required a new way of thinking to realize that it was more fruitful to exploit total internal reflection via an interface rather than brute force reflection from a surface. Although common knowledge now, it wasn't then, and this community-wide groupthink held up fiber development for over 50 years. This paradigm shift occurred in the 1950s, driven by teams in the Netherlands and England. Moller Hausen may have become the first to suggest coating a lower-index material on the exterior of the glass fiber, and he used margarine to coat glass fibers and demonstrate total internal reflection.[3] Van Heel, as well as Hopkins and Kapany, soon experimented and published (in 1954) works describing the use of a different glass for the lower-index medium.[5] Curtis applied for the first patent on glass-clad fibers on May 6, 1957.[4] Moller Hausen went on to invent bubble wrap. Interestingly, from 1955 to

1965, one man was listed as an author on 30 percent of the papers concerning optical fibers.[5] This person was Narinder Kapany, a prodigious researcher and author who was responsible for many innovations and incremental improvements in fiber technology. He founded Optics Technology Inc. in 1961.

Biomedical uses were the first applications of fibers in the 1950s, 1960s, and 1970s, and this kept research alive when few believed that telecom applications would ever be practical. Most new fiber companies drew their venture capital based on the medical markets. Today, fibers are still frequently used for visual exploration of the gastrointestinal tract, inspecting colons and bladders, and for laparoscopies. Although important, medical and imaging uses of fibers proved to be niche markets as compared with telecom.

In 1966, a paper authored by Charles Kao and George Hockham (of Standard Technologies Laboratories), assuming improved purity, calculated that a 1-GHz signal could be sent 10 km over a particular fiber structure, as Rayleigh scattering is only a few decibels per kilometer at 1 micron.[6] The race was on at Corning, Standard Technologies Labs, Telefunken, Nippon Telephone and Telegraph, and eventually (two years later), a low-loss fiber was developed by Bell Labs and others. In 1970, the race seemed to be won by Corning, as Kapron et al. reported a fiber with a 20-dB/km transmission loss. This was high by today's standards, but it was the first time a fiber had been produced that had a chance to work over telecom distances, and everyone knew it. By 1972, there were many papers reporting fiber with less than a 5-dB/km loss. There was a paper published in 1973 by Hasegawa on the theoretical application of solitons to the long distance transmissions in optical fiber, which fortified the notion of optical communication via fibers. The application was experimentally demonstrated in 1980 by Mollenauer et al.[7,8] The first fiber optic trade show, Fiber-Con, was held in 1978. That year also saw fiber-to-the-home deployment in Japan by Nippon's Hi-OVIS project and the announcement of a 0.2-dB/km loss at 1.55 microns.

In the early 1980s, French, British, Japanese, and Canadian telecommunication companies laid fiber optic networks that could handle a few million bits per second. This planted the seed for the explosion of the 1990s and was crystallized in American minds by Sprint's "pin-drop" campaign. In a huge marketing effort, Sprint Telecom advertised that you could hear a pin drop across a continent though their fiber optic network.

In the 1990s, the networks concentrated on gigabit data links and installing the newly invented EDFAs. The proliferation of cell phones and the Internet spurred network companies to continue building multibillion-dollar networks. The stock market rewarded this unbridled expansion. Almost all of the telecom companies went public before they made a dime of profit and saw their stock boom in value like the railroad stock of another infrastructure bubble, over a century earlier. With a looming recession and cries of a fiber glut, the market bubble burst in 2000, the money driving the expansion dried up, fortunes were lost, and engineers were laid off. The imposed austerity forced a more frugal and practical approach to telecommunications development.

This chapter contains several rules relating to fiber physics. Other rules regarding fiber manufacturing, strength, and reliability are covered in a separate chapter, as are the subjects of splicing, connecting fibers, and multiple-fiber cables. A number of rules address the number of modes or what makes a fiber a single-mode one. One rule gives a simple equation to estimate the absorption of a fiber, and other gives the accepted windows and approximate absorption.

There is a rule providing equations and relationships to estimate the mode field radius and diameter. We apologize for using both radius and diameter, but there seems to be general disarray in the industry; sometimes diameter is used, and sometimes radius, so be cautious of that pesky factor of two.

For the reader who is interested in more details concerning optical fibers, several books can be found (such as Kapany's *Fiber Optics*). If the reader is not familiar with fiber optic technology, the authors suggest first reading these classic optics books to gain a familiarity with the basics. Excellent introductory discussions can be found in E. Hecht's *Optics*, E. Hecht's *Fiber Optics*, Hobbs' *Building an Electro-Optical Systems*, McGraw-Hill's *Handbook of Optics* (Vol. 4), and Jeff Hecht's *City of Light*. Additionally, there are numerous conferences each year (e.g., SPIE, IEEE, OFC, NFOEC), which usually have multiple sessions on fiber topics.

References

1. Private communications with A. Girard, 2002.
2. W. Gambling, "The Rise and Rise of Optical Fibers," *IEEE Journal on Selected Topics in Quantum Electrodynamics*, 6(6), p. 1085, Nov./Dec. 2000.
3. J. Hecht, *City of Light*, Oxford University Press, New York, p. 51, 1999.
4. J. Hecht, *City of Light*, Oxford University Press, New York, p. 67, 1999.
5. J. Hecht, *City of Light*, Oxford University Press, New York, p. 70, 1999.
6. W. Gambling, "The Rise and Rise of Optical Fibers," *IEEE Journal on Selected Topics in Quantum Electrodynamics*, 6(6), p. 1086, Nov./Dec. 2000.
7. A. Hasegawa and F. Tappert, "Transmission of Stationary Nonlinear Pulse in Dispersive Dielectric Fibers, I. Anomalous Dispersion," *Appl. Phys. Lett.* 23, pp. 142–144, 1973.
8. L. Mollenauer, P. Stolen, and J. Gordon, "Experimental Observation of Picosecond Pulse Narrowing and Solitons in Optical Fibers," *Phys. Rev. Lett.* 45, p. 1095, 1980.

NUMERICAL APERTURE

Numerical aperture (NA) depends on the index of refraction of the core and the cladding of a fiber.

Discussion

NA is given by

$$NA = \sqrt{n_f^2 - n_c^2}$$

where f = fiber
       c = cladding

NA is also equal to the sine of the half-angle of the cone of light either projected by the fiber or from which it can accept light. Single-mode fibers tend to have an *NA* of about 0.1, while multimode fibers have a higher NA of 0.2 to 0.3, making it easier to launch light into them.

Reference

1. I. Jacobs, "Optical Fiber Communication Technology and System Overview," Chap. 2 in *Fiber Optics Handbook,* M. Bass, Ed. in Chief, and E. Van Stryland, Assoc. Ed., McGraw-Hill, New York, p. 2.2, 2001.

DEPENDENCE OF NA ON FIBER LENGTH

When the numerical aperture (NA) is measured for a long multimode fiber, it is usually found to be less than for a short segment of the identical fiber.

Discussion

This rule is based on waveguide physics and supported by empirical observations. It depends on the state of the art at the turn of the recent century and is generally useful for comparing length differences of factors of two or more.

This should be used to add margin to NA in the design process and is useful as a reminder that the NA will be less for long fibers.

This rule of thumb should be heeded when making measurements or determining specifications. This phenomenon results in the relatively greater attenuation of the high-order (high-angle) modes (or rays) within the fiber.

References

1. The technical Staff of CSELT, *Optical Fiber Communication*, McGraw-Hill, New York, p. 267, 1981.
2. N. Lewis and M. Miller, "Fiber Optic Systems," in Vol. 6, *Active Electro-Optical Systems*, C. Fox, Ed., *The Infrared & Electro-Optical System Handbook*, J. Accetta and D. Shumaker, Exec. Eds., ERIM, Ann Arbor, MI, and SPIE, Bellingham, WA, p. 245, 1993.

SPECTRAL ATTENUATION (FIBER ATTENUATION AS A FUNCTION OF WAVELENGTH)

Attenuation based on Rayleigh scattering is approximately[1]

$$\alpha \approx \frac{0.9}{\lambda^4} \tag{1}$$

where α = attenuation of the fiber, expressed in dB/km
λ = wavelength in microns

Including the molecular IR absorption that occurs at the longer wavelengths, the intrinsic attenuation of a fiber can be expressed as[2]

$$\alpha \approx A\exp\left(\frac{-48.5}{\lambda}\right) + \left(\frac{B}{\lambda^4}\right) \tag{2}$$

where A = assume it to be 7.81×10^{11} dB/km
B = Rayleigh wavelength loss coefficient (given by one reference as 0.9 and another as 0.973)

Discussion

Given modern high-purity fibers, the fiber's transmission is dominated by Rayleigh scattering (due to irregular glass structure) up to about 1.35 microns, where OH absorption then dominates; for wavelengths larger than 1.6 microns, infrared molecular absorption is the dominant absorption factor. Fibers can be now made without the OH peak. At the shorter end of the spectrum, UV attenuation is also a function of the wavelength and increases very quickly with smaller wavelength. Also, bending losses can contribute to losses at the higher wavelengths.

The first equation[1] is for the Rayleigh scattering only, and the 0.9 is the scattering coefficient.

The second equation (from Wilder[2]) is apparently based on an empirical curve fit of the absorption of fibers as a function of wavelength. The equation does a good job of representing the Rayleigh scattering, which dominates to near 1.6 microns, and the molecular absorption, which begins to dominate the absorption at wavelengths longer than about 1.6 microns. It does not include the OH absorption peak near 1.4 microns. The −48.5 depends on the fiber and may change with new fiber technologies. Be sure that the units are in microns to match the given A constant. The 7.81×10^{11} dB/km value of the constant is big (probably from an empirical curve fit), but it works.

The authors of this book plotted these equations in Fig. 9.1 for typical telecom wavelengths. If pressed for an answer in a meeting, the practitioner can use Eq. (1) until about the middle of the C band (1530 to 1565 nm). Beyond that, the scattering is not the dominant absorption mechanism, so pull out your calculator and resort to Eq. (2).

Wilder explains the increase in absorption beyond 1.6 microns as follows:

> As seen in the calculated attenuation figures, the attenuation is beginning to increase rapidly above 1625nm due to infrared absorption. However, the increased intrinsic attenuation is not the complete story in the 1650nm region. At the longer wavelength, the primary mode becomes very loosely coupled to the core of the fiber and is much more susceptible to macro- and microbending losses. Environmental and mechanical fiber loss in the cable have been known to increase the attenuation making these higher wavelengths unusable for testing on some fibers.[2]

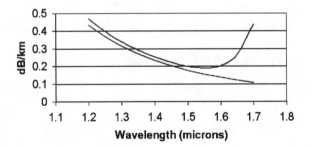

Figure 9.1   Fiber attenuation.

References

1. I. Jacobs, "Optical Fiber Communication Technology and System Overview," in *Handbook of Optics*, Vol. 4, M. Bass, Ed. in Chief, McGraw-Hill, New York, pp. 2.2–2.3, 2001.
2. M. Wilder, "Testing Fibers Carrying DWDM Traffic," *Proceedings of the National Fiber Optic Engineers Conference*, 1999.

BENEFITS OF SMALL NUMERICAL APERTURE

1. Lower numerical aperture (NA) gives more bandwidth, but it is harder to couple power into the fiber.
2. The smaller the core, the lower the attenuation. The smaller the core, the higher the bandwidth.
3. The smaller the core, the lower the cable cost. Small cores may lead to increased prices in other areas, e.g., connectors.

Discussion

The performance characteristics of fiber cables vary depending on the materials used and the process of manufacturing. Moreover, the above basic rules underscore why there is a trend toward reducing fiber size in cables in addition to merely using single-mode fibers. As newer technologies emerge that reduce the cost and complexity of using a certain smaller size, the cables become cost effective, and the next reduction in size awaits the next connection or network element breakthrough. For additional details, see the Numerical Aperture rule (p. 220).

Reference

1. http://www.kei-telecom.com/strategies/fiber_cable_construction.htm, 2002.

TRANSMITTED FIBER SPOT SIZE

A laser spot's size at the end of the fiber, after transmission though a fiber, can be calculated as

$$\omega = b\frac{a}{\sqrt{\ln V}}$$

where ω = spot size

b = a fudging constant to account for real-world misalignments and unaccounted diffractive effects; generally between 1.02 and 1.1 for well aligned systems

a = fiber core radius

V = V parameter, defined as $V = kan_1\sqrt{2\Delta}$, where n_1 is the core index of refraction, k is the propagation constant $(2\pi/\lambda)$, and Δ is the relative refractive index difference defined on p. 227

Discussion

A laser's spot size can be assumed to be Gaussian, and thus the calculation can be simplified to the above equation. The authors of this book added the b factor to the equation from the source to account for real-world conditions. The Gaussian is about the best one can do, so if "1" is used for b, that would indicate the "best-case" scenario. For a well designed and aligned system, it really shouldn't be above ≈1.1.

It is important for the designer to realize that the spot size obviously is a function of wavelength, and this manifests itself in the above equation in the k factor and in the index of the core (which slightly changes as a function of wavelength). Spots in the L band are larger than those in the C band for the same system. This means that the accuracy of an optical signal analyzer (OSA) or optical performance monitor (OPM) often will be worse at the longer wavelengths. Moreover, spot sizes for shutters, optics, couplers, receivers, and other optical components should be calculated at the longest expected wavelength.

Figure 9.2 illustrates this. As one can see, the spot size is larger at longer wavelengths. Moromoto et al. explain that the figure

...shows the calculation results of the spot size of the optical beam, ω against wavelength using a = 4.2 μm and Δ = 0.35 percent with the wavelength range between 1530 nm and 1580 nm, which corresponds to C-band in DWDM transmission systems. An experimental result is also shown (as the higher line). Though the absolute values between calculation and experiment are slightly different, it is clearly shown that the spot size of the optical beam of 1530 nm is about 0.2 μm smaller than that of 1580 nm, therefore, it is predicted

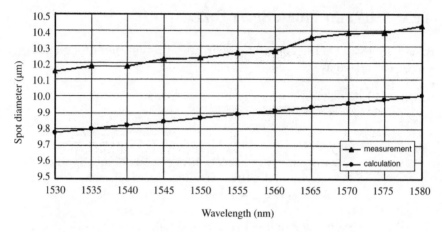

Figure 9.2 Spot diameter vs. wavelength.

that the attenuation value at a particular shutter position depends on the wavelength, since a beam spot size varies with wavelength.[1]

Reference

1. M. Moromoto, "Improvement in Optical Performance of MEMS-Based Variable Optical Attenuator," *Proceedings of the National Fiber Optic Engineers Conference,* pp. 932–933, 2001.

GOOD FIBER MODE VOLUME

Good guiding and reliable single-mode operation occurs for normalized frequency (V) values from about 1.8 to 2.4.

Discussion

The normalized frequency, V, is defined as

$$V = kan_1\sqrt{2\Delta}$$

where $\quad k_o = 2\pi/\lambda$
$a =$ core radius
$n_1 =$ core index
$\Delta =$ normalized index difference, defined as $(n_1^2 - n_2^2)/(2n_1^2)$
where n_2 is the cladding index

As Hobbs explains, light is not confined entirely to the core but actually spreads into the cladding. The fiber's ability to guide the light becomes weaker and weaker as more light enters the cladding, leading to a practical minimum for the core diameter. When the mode volume ranges from 1.8 to 2.4, high-quality single-mode operation is possible.

Hobbs also points out, "When we talk about a fiber being single mode, we are of course talking about the scalar approximation, whereas the light in the fiber has two degrees of freedom in polarization."

The reader should also see the other rules concerning mode field diameter.

Reference

1. P. Hobbs, *Building Electro-Optical Systems, Making It All Work,* John Wiley & Sons, New York, pp. 240–245, 2000.

MODE FIELD RADIUS AND DIAMETER

The mode field diameter (MFD) of a single-mode fiber is roughly 6 to 10 times the operating wavelength, and sometimes as much as 20.

Discussion

The mode field radius is approximately Gaussian, and $1/e^2$ mode field radius for single-mode fibers is approximated by Hobbs as[3]

$$\omega_0 \approx a\left(\frac{1}{3} + \sqrt{\frac{2.6}{V^3}}\right)$$

where ω_0 = mode field radius
$\quad\quad a$ = core radius
$\quad\quad V$ = normalized frequency (see associated rule)

Additionally, Barzos gives the following approximate equation for the fiber mode radius:[4]

$$\omega_0 \approx \frac{d_f(0.65 + 1.619\,V^{-1.5} + 2.879\,V^{-6})}{2}$$

where d_f = diameter of the fiber's core

This rule is based on the physics of light propagation in a fiber waveguide. The mode field diameter (MFD) depends on several parameters, including numerical aperture (which decreases the MFD) and wavelength (longer wavelengths lead to larger MFDs).

The MFD is needed to calculate the coupling efficiency from another fiber, a laser source, or a detector. This is also useful to remember when sizing a fiber.

It is desirable for the MFD to be as large as possible so the maximal light can be coupled into the fiber. However, it should not be so large that higher-order modes can propagate. The physics of light propagation determines the beam diameter.

The core of a fiber can be physically measured due to its higher index of refraction. In a multimode fiber, the light is confined to the core through total internal reflection. In a single-mode fiber, the light has a distribution that approximates a Gaussian (actually, a Bessel function in the core and a separate decaying exponential in the cladding). The mode field diameter is the point at which an equivalent Gaussian distribution has decayed to its $1/e^2$ points and is approximately 15 percent larger than the core diameter.

Some fibers, such as the Corning "Pure Mode HI 980," have a much smaller mode field diameter.

References

1. Private communications with G. Ronan, 1995.
2. Product catalog, Wave Optics Inc., Mountain View, CA, p. 23, 1995.
3. P. Hobbs, *Building Electro-Optical Systems, Making It All Work,* John Wiley & Sons, New York, p. 245, 2000.
4. Barzos et al., "Parameter Extraction for a Figure of Merit Approach for the Comparison of Optical Interconnects," *Optical Engineering,* 37(10), p. 2775, Oct. 1998.
5. H. Farley et al., "Optical Fiber Analysis Using Moving Slits, CCD Arrays, and Goniometer Techniques," *Proceedings of the National Fiber Optic Engineers Conference,* 2000.

BANDWIDTH OF A MULTIMODE FIBER

1. The bandwidth of a multimode fiber can be estimated as

$$BW_{eff} = \frac{BW_1}{L^\gamma}$$

where BW_{eff} = effective bandwidth at the end of the fiber in MHz-km
BW_1 = modal bandwidth of 1 km of fiber
L = length of the fiber, generally in kilometers
γ = an empirical scaling factor, generally between 0.5 and 1

2. Conversely, the maximum data rate that can be transmitted down a multimode fiber can be approximated as

$$DR_{max} \approx \frac{1}{4\Delta\tau}$$

where DR_{max} = maximum data rate
$\Delta\tau$ = modal dispersion

Discussion

The problem with modal dispersion is that the many modes do not travel at the same velocity in the fiber and consequently do not arrive at the end of the fiber at the same time. This causes a broadening and a power loss of the pulse composed of these modes.

This rule applies only to multimode fibers and gives an estimate of the impact of mode mixing on the bandwidth. Although most emphasis at the turn of the twenty-first century was to move to single-mode fibers for long distance optical telecom to eliminate multimode pulse dispersion, some believe that a different paradigm may be once again be considered if the pulse dispersion problem can be solved. Regardless of the dispersion, multimode fibers are frequently used in lengths of under 300 m to support data rates of over 3 Gb/s. These low-cost fibers are especially useful in enterprise networks and in buildings housing telecom equipment.

A multimode fiber can propagate many modes (e.g., thousands for thick fibers), each of which can carry data, but they also can interact with one another. When light is launched into a multimode fiber, the rays enter at different angles. As they traverse the fiber, the modes can interfere, which is referred to as *mode mixing*. This will result in a limiting of the bandwidth as described above.

Marusarz and Sayeh[2] suggest that a neural network might be employed to unscramble the received data. Equation 2 from this reference assumes analog transmission of data and successfully demultiplexing and identifying the received data, and it considers modal dispersion. Marusarz and Sayeh point out that a method for calculating the number of modes is approximated by

$$N_m \approx \frac{V^2}{2}$$

where N_m = number of modes
  V = normalized frequency, which can be defined in a slightly different form from that shown earlier,

$$V = \frac{2\pi a\, NA}{\lambda}$$

where a = fiber radius
  NA = numerical aperture
  λ = free-space laser wavelength

Thus, with the above in mind, the modal dispersion ($\Delta\tau$) can be calculated as

$$\Delta\tau = \frac{L(n_1 - n_2)}{c}\left(1 - \frac{\pi}{V}\right)$$

where L = fiber length
  c = speed of light
  n_1 = index of refraction of the core
  n_2 = index of refraction of the cladding

The maximum data rate can also be written as

$$DR_{max} = \frac{0.25}{\left[\frac{L(n_1 - n_2)}{c}\left(1 - \frac{\pi}{\sqrt{2N_m}}\right)\right]}$$

where the variables are as defined previously.

As the number of modes is increased, R. Marusarz and M. Sayeh point out that the maximum data rate decreases to a minimum at

$$\frac{0.25\,c}{[L(n_1 - n_2)]}$$

"However, as the number of modes is increased by use of a larger-diameter cable, the potential data rate increases because more channels (modes) are available for parallel data transmission. This analysis suggests the benefits that can be achieved when larger-diameter multimode fiber-optics cables are used to enhance data-handling capability."[2]

References

1. S. Kartalopoulos, *Introduction to DWDM Technology*, IEEE Press, Piscataway, NJ, p. 47, 2000.
2. R. Marusarz and M. Sayeh, "Neural Network-Based Multimode Fiber-Optic Information Transmission," *Applied Optics* 40(2), pp. 220–221, 2001.

WAVEGUIDE RULES

1. In a dielectric waveguide, all but the cylindrical symmetric modes (TE_{om} and TM_{om}) are hybrid modes.
2. The number of modes increases with the square of the fiber diameter.
3. Almost all modes have cutoffs that are a function of wavelength, fiber diameter, and fiber numerical aperture.
4. The cutoff values for various modes are given by a characteristic equation,

$$U_{nm} = d(NA)/\lambda$$

where U_{nm} = mth root of a cut-off condition involving nth-order Bessel function
d = diameter of the fiber
NA = numerical aperture of the fiber
λ = wavelength of the light

Discussion

In optical communication fibers, the waveguide properties can become important, and single-mode fibers are sometimes required. These rules are based on simple physics, approximations to theory, and observations of the current state of the art, and they provide some simple relationships about the waveguide properties.

These are valid for small fibers, where the transverse dimensions of the fibers approach the wavelength of the light. Therefore, this is especially of concern for infrared fibers.

The permissible modes within a fiber can be a critical design parameter and limit the transmission data rate. If the fiber's size is close to that of the wavelength of light being propagated, only specific distributions of the electromagnetic field will satisfy Maxwell's equations and thereby be allowable. This results in discrete waveguide modes. In large fibers, there are many modes that interact (usually to the detriment of the engineer) and overlap. Therefore, the above waveguide properties are often inaccurate for large fibers.

The cutoff wavelength is the wavelength under which the fiber will become multimode; if larger, it will be single-mode.

References

1. N. Kapany and J. Burke, *Journal of the Optical Society of America,* 51, pp. 1067–1078, 1961.
2. E. Snitzer, "Cylindrical Dielectric Waveguide Modes," *Journal of the Optical Society of America,* 51, pp. 491–498, 1961.
3. W. Siegmond, "Fiber Optics," in W. Driscoll, and W. Vaughan, Eds., *Handbook of Optics,* McGraw-Hill, New York, pp. 13–11, 1978.

NUMERICAL APERTURE DETERMINES TOTAL INTERNAL REFLECTION

$$NA = \sin A = (n_1^2 - n_2^2)^{1/2}$$

where NA = nominal numerical aperture
 A = entrance (or acceptance) half-angle (radians)
 n_1 = index of refraction of the fiber medium
 n_2 = index of refraction of the fiber cladding (always lower for total internal reflection)

Discussion

This is useful in the design stages to estimate the acceptance angle of a fiber. Conversely, if one has a given fiber, this equation can indicate the acceptance angle for the launch. This rule is based on determining the numerical aperture of a fiber from Snell's Law, basic trigonometry, and fiber optic theory. Although it is frequently applied for rays across the fiber optic edge, it really is derived for the meridional ray (or central ray) only.

Since the above equation assumes that the index of the air is 1, light must enter from a medium with an index of refraction close to 1 (e.g. air or space but not a high-index immersion oil).

The actual acceptance angle is usually not as sharply defined, as the rule contends, due to diffraction, striae, and surface irregularities.

A ray that encounters the exterior fiber face with an angle of less than or equal to the acceptance angle will undergo total internal reflection when it encounters the difference in index of refractions between the cladding and the fiber media. The numerical aperture can be "tuned" for a larger NA by making the difference between the core and cladding greater. But beware, because, in a single-mode fiber, the NA will get larger as the core size becomes smaller.

OPTICAL FIBER BANDPASSES

A fused-silica optical fiber will transmit wavelengths between 250 and 2000 nm (0.25 and 2 µm) but is limited to specific wavelength windows for long-range transmission.

Discussion

The long-range transmission windows are typically defined as shown in the following table:

Band nomenclature	Name	Approximate absorption (dB/km)	Wavelength (nm)
Fiber channel	Fiber channel	1.9	770–860
IEEE serial bus	IEEE serial bus	1.9	830–860
O	Original	0.29	1260–1360
E	Extended	0.19	1360–1460
S	Short wavelength	0.22	1460–1530
C	Conventional	0.2	1530–1565
L	Long	0.23	1565–1625
U	Ultra-long	0.29	1625–1675

Most telecom fibers are made from fused silica glass, and nature decreed that this material will transmit only a limited range of wavelengths. Attenuation within a fiber is caused by material absorption, scattering, waveguide attenuation, and leaky modes. Material absorption is caused by mechanical resonance in the crystalline structure of the waveguide material and by absorption peaks resulting from metal and water impurities in the glass.

Usually, the "glass" used in optical fibers is quartz (silica) and must be ultra-pure for low-loss transmission. Wavelengths shorter than 250 nm are attenuated because of absorption and Rayleigh scattering, and those above 2 µm are severely attenuated due to infrared resonant absorption bands within the material. It was in the 1960s when processes became advanced enough to routinely remove enough impurities to create a fiber with an attenuation of <20 dB/km. By contrast, the attenuation of common glass window material is much greater than 1000 dB/km. With the exception of the water resonance absorption (at 1240, 1280, and 1380 nm), the absorption closely follows that expected from Rayleigh scattering to about 1.55 microns. The increase in absorption beyond 1.55 microns is due to atomic resonances. See the related rule, "Spectral Attenuation," on p. 222.

This rule applied to fibers, not hollow waveguides or fibers of special materials. There are UV and infrared fibers (e.g., ZBLAN), but they are generally not used for optical telecommunication at the present time. Additionally, small concentrations of dopants (such as germanium, phosphorous, boron, or fluoride) are sometimes added to change the index of refraction and may alter the transmission properties.

The definition of spectral bands is to facilitate discussion and is not for specification. The specifications of operating wavelength bands are given in the appropriate system recommendations. The G.65x Fiber Recommendations have not confirmed the applicability of all these wavelength bands for system operation or maintenance purposes. Additionally, the boundary (1460 nm) between the E band and the S band continues under study.[4]

The U band is currently reserved for maintenance purposes only, and transmission of traffic-bearing signals is not currently foreseen. The use for nontransmission purposes must be done on a basis of causing negligible interference to transmission signals in other bands.[4]

The above table shows regions of low attenuation where optical sources and detectors are readily available. Typical fibers have very low attenuation in these regions, and they avoid spectral regions of high scatter or hydroxyl absorption. It is anticipated that, in the near future, various applications, with and without optical amplifiers, will utilize signal transmission covering the full range of 1260 to 1625 nm. As the technology progresses, additional windows (more to the infrared) are likely to emerge.

References

1. Private communications with G. Ronan, 1995.
2. M. Wilder, "Testing Fibers Carrying DWDM Traffic," *Proceedings of the National Fiber Optic Engineers Conference,* 1999.
3. I. Jacobs, "Optical Fiber Communication Technology and System Overview," in *Handbook of Optics,* Vol. 4, M. Bass, Ed. in Chief, McGraw-Hill, New York, pp. 2.2–2.3, 2001.
4. Private communications with A. Girard, 2002.

SINGLE-MODE CONDITION RULES

1. The cutoff wavelength for single-mode operation is captured in the following equation

$$d = 0.767\lambda_c/NA \qquad (1)$$

where d = core diameter in microns
λ_c = cutoff wavelength in microns
NA = numerical aperture

2. The following calculation defines the wavelength above which only the single (or lowest order axial) mode exists:

$$\lambda_c = \frac{2\pi}{2.405}\, an_1\sqrt{2\Delta} \qquad (2)$$

where a = core radius, generally 1.5 to 5 microns for single-mode fibers
n_1 = index of refraction of the core, generally 1.47 to 1.52
Δ = normalized index difference between the core and cladding, generally on the order of 0.003 to 0.01 for a single-mode fiber. It is generally defined as on p. 227 but can be approximated by

$$\frac{n_{core} - n_{clad}}{n_{core}}$$

where n_{clad} and n_{core} = indices of refraction of the cladding and core of the fiber, respectively

3. Another form is

$$\frac{2\pi a(NA)}{\lambda} < 2.405 \qquad (3)$$

Discussion

The above are all different mathematical versions of the same physics concept. These equations give the relationship of core diameter, numerical aperture, and cutoff wavelength of a single-mode optical fiber.

A single-mode fiber is one in which only a single mode of the light's electromagnetic fields can propagate. Single-mode fibers are the dominant fiber in telecom. These fibers allow the efficient transmission of a single mode of laser without multiple modes mixing and interfering. Moreover, different modes travel at different speeds, leading to blurring of the signal, especially for high-bandwidth communication. For single-mode operation, a fiber needs to have a radius on the order of its wavelength of operation or a small numerical aperture. When launching light into a fiber, a fiber has only a tiny contribution to field s in the z (length of the fiber) direction, so the launched signal is analogous to a TEM wave. The condition for single-mode propagation is that the normalized frequency be less than the above condition [defined in Eq. (3)]. As such, single-mode fibers for optical telecommunication typically have a core diameter between 5 and 10 microns.

For typical C- and L-band operation, 1.625 microns is the usual cutoff. Thus, Eq. (1) above can be further simplified to

$$d = 1.25/NA$$

where d and λ are in the same units.

This illustrates why single-mode fibers for optical telecom have such small core diameters. Please note the other rule in this chapter on numerical aperture (p. 220).

Hobbs points out that good single-mode operation occurs when the normalized frequency is in the range of 1.8 to 2.4, as stated in another rule in this chapter.

Stated another way, when the core diameter, wavelength, and indices of the core and cladding have a specific range of values, only a single mode will propagate in the fiber, or

$$V = \frac{2\pi a}{\lambda} \sqrt{n_1^2 - n_2^2}$$

Here, a is the fiber core radius and the subscripts refer to (1) core and (2) cladding. To achieve single-mode propagation at typical telecom wavelengths, the difference between the indices is less than 0.3 percent, and the core has a diameter of about 9 microns.

References

1. A. Cook, from S. Weiss, "Rules of Thumb, Shortcuts Slash Photonics Design and Development Time," *Photonics Spectra*, Oct. 1988, pp. 136–139.
2. P. Hobbs, *Building Electro-Optical Systems, Making It All Work*, John Wiley & Sons, New York, p. 245, 2000.
3. I. Jacobs, "Optical Fiber Communication Technology and System Overview," in *Handbook of Optics*, Vol. 4, M. Bass, Ed. in Chief, McGraw-Hill, New York, pp. 2.2–2.3, 2001.
4. N. Lewis and M. Miller, "Fiber Optic Systems," in Vol. 6, *Active Electro-Optical Systems*, C. Fox, Ed., *The Infrared and Electro Optical Systems Handbook*, J. Accetta and D. Shumaker, Ex. Eds., ERIM, Ann Arbor, MI, and SPIE, Bellingham, WA, p. 248, 1993.
5. I. Jacobs, "Optical Fiber Communication Technology and System Overview," Chap. 2 in M. Bass, Ed. in Chief, and E. Van Stryland, Assoc. Ed., *Fiber Optics Handbook*, McGraw-Hill, New York, p. 2.2, 2001.
6. http://www.ofr.com/tools, 2002.

SIZING A FIBER

Usually, the core diameter of a multimode fiber ranges from 50 to 1000 μm. Also, a fiber's outer diameter is typically 1 to 20 times the core diameter.

Discussion

This rule is based on currently produced fibers. The state of the art changes, and the desire is for smaller fiber diameters. Also, custom-made fibers can be of much different sizes and ratios, and at least one of the standard sizes does not fit this rule. The above is useful for estimating available fiber sizes and understanding the expected size of a fiber.

Dimensions of step-index fibers vary greatly (see Fig. 9.3), depending on the intended application. Various fiber manufactures have developed standard sizes. The above rule is based on fibers that were available as this was written, and the small diameters do not necessarily follow this rule.

Standard sizes of core and cladding include:

- 8- to 9-μm core with a 125-μm cladding (and a 1300-nm single mode)
- 62.5/125 for the Fiber Data Distribution Interface (FDDI) standard
- 85/125, the typical European standard
- 100/140, the typical U.S. military standard

Also, for comparison, the Corning SMF-28 has a 9-micron core and a 125-micron outer diameter, and their HI1060 has a 5-micron core and a 125-micron diameter.

The sizes that are single mode are not just size dependent—the indices of the core and cladding are important, too.

References

1. Private communications with Judy McFadden, 1995.
2. N. Lewis and M. Miller, "Fiber Optic Systems," in Vol. 6, *Active Electro-Optical Systems*, C. Fox, Ed., *The Infrared and Electro Optical Systems Handbook*, J. Accetta and D. Shumaker, Ex. Eds., ERIM, Ann Arbor, MI, and SPIE, Bellingham, WA, p. 245, 1993.

Cladding with an index of refraction of $>n_1$

Core with an index of refraction of n_1

Protective coating

Figure 9.3   Step-index fiber.

NUMBER OF MODES IN A FIBER

The number of modes in a step index fiber is[1]

$$N_s = (1/2) V^2$$

and the number of modes in a graded index fiber is[1]

$$N_g = (1/4) V^2$$

where N_s = number of modes in a step index fiber
V = normalized frequency (see associated rules in this chapter)
N_g = number of modes in a gradient index fiber

Discussion

The propagation of data down a fiber depends on the purity of the fiber and its cladding as well as the cross-sectional dimensions. Fibers with core diameters beyond 25 microns are generally considered multimode fibers. The equations define when the size and the indices restrict the fiber to a single mode. In the application of these fibers, it is often useful to know how many modes of light transmission will propagate down these fibers.

In multimode operation of an optical fiber having a power-law refractive index profile, the approximate number of bound modes, i.e., the mode volume, is given by

$$\frac{V^2}{2} \left(\frac{g}{g+2} \right)$$

where V = normalized frequency greater than 5
g = profile parameter

The profile parameter (g) defines the shape of the refractive-index profile. *Note:* The optimal value of g for minimal dispersion is approximately 2.[3]

Note that a step index fiber supports twice as many modes as a graded index fiber of the same core radius and numerical aperture operated at the same wavelength.

References

1. http://www.fontcanada.com/gloss.html.
2. S. Kartalopoulos, *Introduction to DWDM Technology,* IEEE Press, Piscataway, NJ, p. 42, 2000.
3. http://www.its.bldrdoc.gov/fs-1037/dir-024/_3599.htm.

RADIATION DAMAGE

Fibers with a pure silica core are quite insensitive to common radiation. Optimizing fiber performance by adding dopants makes the fiber susceptible to radiation effects.

Discussion

Dopants are used to control the index of refraction of the core and cladding and to manage dispersion properties. These dopants make the fiber susceptible to both radiation-induced darkening and scintillation in the fiber. Scintillation has the effect of creating noise in the fiber with an adverse impact on bit error rate.

A simple model of the increased bit error rate (BER) after radiation exposure is

$$BER = BER_o + A \times dose^b$$

where A and b are empirically fitted. BER_o is the bit error rate before irradiation, and *dose* is in rads.

Reference

1. C. DeCusatis and G. Li, "Fiber-Optic Communication Links (Telecom, Datacom, and Analog)," Chap. 6 in *Fiber Optics Handbook,* M. Bass, Ed. in Chief, and E. Van Stryland, Assoc. Ed., McGraw-Hill, New York, p. 6.18, 2001.

Optical Signal Degradation

The authors decided to collect a slew of rules, in a single chapter, devoted to the degradation of the optical signal. Many of these are fiber related as they deal with the propagation of light down a long fiber. These rules deal with the purely optical effects of dispersion, four-wave mixing, and other nonlinear effects that occur when sending an optical pulse down a length of fiber. Thus, these rules do not address the important issue of amplitude noise generation, which is covered in Chap. 8. They also do not speak to signal degradation as a result of connectors, switches, amplifiers, or other network elements.

The phenomena described in this chapter manifest themselves as an increase in the length of a data pulse or a decrease in optical signal or in signal quality, and/or the generation of noise. The effects we include may affect the signal-to-noise ratio (SNR) of the system. They can also appear as an increase in noise at one wavelength due to the presence of a signal in another channel. An example would be four-wave mixing of dense wavelength division multiplexing (DWDM) signals. Temporal lengthening of a pulse as a result of *differential group delay (DGD)* may cause one pulse to mix with its neighbor, affecting the basic SNR.

Dispersion results when there are different speeds in the fiber for different modes. In multimode fibers, both chromatic and intramodal dispersion (different propagation speeds for different modes) can occur. This phenomenon generally limits the length of high-data-rate multimode fibers to a few hundred meters or less. Single-mode fibers exhibit only chromatic dispersion and intramodal dispersion (dispersion within a single mode). Different polarization modes travel at different velocities inside a single pulse or among different pulses of a DWDM systems, and they can interfere with some phase-related nonlinear effects such as self-phase or cross-phase modulation.

Most of the rules in this chapter concern the effects of various types of these dispersions. Some are restatements of the same principle with a new simplification or interesting shortcut applied to it, or a rule may be a simplification for a particular application. Many of these rules are closely related to each other and occasionally related to a rule in Chap. 8, "Noise." Some even have a close analog in the atmosphere as addressed in Chap. 5, "Free-Space Optical Communications."

With the advent of 40 Gb/s transmission rates, polarization-mode dispersion (PMD) became a key dispersion mechanism. This occurs because the glass material that composes a fiber has a slightly different index of refraction for different polarizations. Therefore, the ordinary rays travel at a slightly different speed from that of the extraordinary rays, and they thus arrive at their destination at slightly different times. These slight differences manifest themselves as the pulse being stretched in length and time—or "broadened." Obviously, if the pulse is broadened to such an extent that it overlaps and interferes with the following pulse, data is corrupted, and failures result.

Some manufacturers have gone to great lengths (pun intended) to decrease a fiber's dispersion at a particular wavelength, commonly at 1310 or 1550 nm. Generally, these are referred to as *dispersion-shifted* fibers, as they have near zero dispersion at a given wavelength but tend to suffer high dispersion elsewhere, so they have limited use for DWDM applications. They also tend to have other deleterious effects such as crosstalk and interference between various spectral channels and susceptibility to certain types of nonlinear effects. Some manufactures have introduced a *dispersion-optimized fiber*, which reduces these negative effects and provides near zero dispersion over a wide spectrum (such as from 1300 to 1560 nm), generally at the cost of increased absorption. Another genre is the *dispersion compensated fiber*, which essentially has negative dispersion profile so that a link of this fiber can compensate for the dispersion resulting from a link of a regular fiber. Adjoining two links of equal length will have the resultant effect of zero dispersion (the first link disperses the signal, and the second effectively "un-disperses" it by dispersing it in the opposite direction). In addition to dispersion caused by polarization/index effects, there is also dispersion caused by chromatic effects and an interesting phenomenon called *four-wave mixing*, where interference can jump to other wavelengths.

The authors find it interesting to note that these effects all are present in the glass and optics of any refractive telescope or imaging system. However, the effects are either so incredibly minute due to the small length of material the light is traversing (several centimeters for an optical telescope as compared to tens of kilometers for a fiber) or because the continuous incoming radiation (as opposed to pulses) makes these issues inconsequential.

NUMBER OF NEW WAVELENGTHS DUE TO FOUR-WAVE MIXING

The number of new wavelengths generated by four-wave mixing is

$$\lambda_n = \frac{N^2(N-1)}{2}$$

where λ_n = number of new wavelengths
   N = number of original data-carrying wavelengths

Discussion

When more than one wavelength is launched down a fiber, an intensity modulation occurs at a beat frequency between the laser wavelengths. This beat frequency results in the creation of sidebands (or lobes) of lower intensity. Two laser lines will result in two new lines. A wavelength division multiplexing (WDM) of 16 laser lines yield the creation of 1920 of these new annoying lines, and a mere 40 laser lines carrying data result in 31,200 new noise wavelengths. The number can become staggering for DWDM, and it simply results in an almost constant background noise source that cannot be avoided. For example, 160 laser lines result in more than 2 million new lines, and 320 laser lines create more than 16 million. This will manifest itself as a general increase in the noise floor.

The smaller the effective area of a fiber, the more this is an issue, because the amplitude will be stronger, so this is more pronounced with the single-mode fibers popularly used in telecom. Large-diameter fibers usually do not exhibit this phenomenon or do so only very weakly.

TYPICAL CHROMATIC DISPERSION

The chromatic dispersion specification of typical fibers is ≈17 ps/nm/km.

Discussion

Chromatic dispersion depends on the operating wavelength of the system, the slope of the dispersion curve (in ps/nm²/km), and the wavelength at which the dispersion is 0. If the fiber has 0 dispersion at around 1310 nm and has a slope of 0.09 ps/nm²/km, then, at an operating wavelength of 1550 nm, the dispersion is

$$D = \frac{S}{4}\left(\lambda - \frac{\lambda^4}{\lambda_0^3}\right) = 17.08 \ \text{ps/nm/km}$$

The formula derives from the derivative of the group velocity with respect to wavelength. Here, S refers to the slope of the dispersion curve (in ps/nm²/km), and the subscript 0 refers to the wavelength for zero dispersion.

The resulting number of 17 represents the typical requirement, above which the fiber will have failed. Some would argue that most legacy fibers have a much lower dispersion, as this was the failure specification, and all of the fiber deployed was actually better than this figure.

References

1. http://www.stanford.edu/class/ee247/HO13.pdf.
2. C. DeCusatis and G. Li, "Fiber-Optic Communication Links (Telecom, Datacom, and Analog)," Chap. 6 in *Fiber Optics Handbook,* M. Bass, Ed. in Chief, and E. Van Stryland, Assoc. Ed., McGraw-Hill, New York, p. 6.11, 2001.
3. A. Willner and Y. Xie, "Wavelength Domain Multiplexed (WDM) Fiber-Optic Communication Networks," Chap. 13 in *Fiber Optics Handbook,* M. Bass, Ed. in Chief, and E. Van Stryland, Assoc. Ed., McGraw-Hill, New York, p. 13.5, 2001.

ALLOWABLE CHROMATIC (OR WAVELENGTH) DISPERSION

The maximum allowable accumulated dispersion for a 1-dB penalty in signal can be expressed as a function of bit rate and written as

$$DL(\text{ps/nm}) < 10^5/B^2$$

where DL = accumulated dispersion (product of dispersion D(ps/nm/km)
            and length L (km)
        B = bit rate in Gb/s

Also, transmission distance (L_D) is limited by chromatic dispersion and the squared inverse of the bandwidth (B).

$$L_D = \frac{1}{BD\Delta\lambda}$$

Discussion

The first equation allows one to quickly estimate the total dispersion allowable for a given data rate and an acceptable 1-dB loss. It simply relates the data rate to the total allowable accumulated dispersion. Note that since allowable dispersion is sensitive to the inverse square of the data rate, going from a 10-Gb/s rate to a 40-Gb/s rate requires 16 times less dispersion. Dispersion management is the balancing of positive and negative dispersion across the optical spectrum of the transmission. The longer the link, the higher the data rate and the wider the optical spectrum, the harder it is to achieve acceptable dispersion.

The maximum analog bandwidth (or the bit rate) at the receiver is typically a strong function on the total dispersion in the total DWDM system. The chromatic dispersion penalty can be expressed as a function of the data rate squared per the above equation. It is important to note that the allowable dispersion is a function of the data rate squared, not just a linear function of the data rate. Some typical values are given in the table below.

Data rate (Gb/s)	Allowable dispersion (ps/nm)
2.5	12,000–16,000
10	800–1,000
40	60–100

This can also be expressed as a function of fiber length for a given fiber dispersion. As a result, higher data rates cannot travel nearly as far as lower ones.

Note that the second equation is correct without B being squared, since the spectral bandwidth ($\Delta\lambda$) is also proportional to B, hence the inverse square behavior. Beyond the distance of the second equation, the pulse broadening exceeds the size of the pulses. A typical example has B = 2.5 Gb/s, dispersion (D) of 14.68 ps/nm/km, and bandwidth of 0.025 nm, limiting the system to 1090 km. An increase of bandwidth to 10Gb/s limits the range to 60 to 70 km. The selection of 14.68 ps/nm/km, of course, depends on the wavelength of operation of the system. A typical fiber might have a wavelength of zero dispersion around 1300 nm with a slope of 0.075 ps/nm^2/km. If we operate the system at 1550 nm, we get a chromatic dispersion of 14.68 ps/nm/km.[5]

References

1. Y. Danzigner, and D. Askegard, "Full Band 40 Gb/s Long-Haul Transport Enabled by High-Order-Mode Dispersion Management," *Proceedings of the National Fiber Optic Engineers Conference*, 2000.
2. C. DeCusatis and G. Li, "Fiber Optic Communication Links (Telecom, Datacom and Analog)," in *Handbook of Optics*, Vol. 4, M. Bass, Ed. in Chief, McGraw-Hill, New York, pp. 6.9–6.14, 2001.
3. Y. Danzigner et al., "Overcoming Chromatic Dispersion and Nonlinear Effects with a New Dispersion Management Device," *Proceedings of the National Fiber Optic Engineers Conference*, 1999.
4. Y. Danzigner and D. Askegard, "High-Order-Mode Fiber," *Optical Networks* 2(1), Jan./Feb., pp. 42–43, 2001.
5. A. Willner and Y. Xie, "Wavelength Domain Multiplexed (WDM) Fiber-Optic Communication Networks," Chap. 13 in M. Bass, Ed. in Chief, and E. Van Stryland, Assoc. Ed., *Fiber Optics Handbook*, McGraw-Hill, New York, p. 13.5, 2001.
6. http://www.Corning.com/photonictechnologies, 2002.

CHROMATIC DISPERSION OF MULTIMODE FIBERS

Chromatic dispersion of multimode fibers depends inversely on the length of the fiber.

Discussion

The following equations provide a method for estimating the chromatic bandwidth of multimode fibers. Here, λ_c is the center wavelength of the transmitter, L is the length of the cable in kilometers, and λ_w is the bandwidth of the source in nanometers (full width, half maximum). For example, the chromatic bandwidth of a 62.5-micron fiber is given as

$$BW_{chrom} = \frac{10^4 L^{-0.69}}{\sqrt{\lambda_w}(1.1 + 0.0189|\lambda_c - 1370|)}$$

and the similar equation for a 50-micron fiber is

$$BW_{chrom} = \frac{10^4 L^{-0.65}}{\sqrt{\lambda_w}(1.01 + 0.0177|\lambda_c - 1330|)}$$

Clearly, the two equations apply to different operating conditions. Both provide bandwidth in megahertz.

The total bandwidth of a multimode fiber combines the effects of chromatic and modal dispersion according to

$$\frac{1}{BW^2} = \frac{1}{BW^2_{chrom}} + \frac{1}{BW^2_{modal}} + \frac{1}{BW^2_{polar}}$$

This limit on the bandwidth of the fiber can be overcome by using more power at the transmitter. The power penalty is

$$P_D = 1.22 \left[\frac{\text{Rate (Mb/s)}}{BW_{total}\,(\text{MHz})} \right]^2$$

Reference

1. C. DeCusatis and G. Li, "Fiber-Optic Communication Links (Telecom, Datacom, and Analog)," Chap. 6 in *Fiber Optics Handbook*, M. Bass, Ed. in Chief, and E. Van Stryland, Assoc. Ed., McGraw-Hill, New York, p. 6.10, 2001.

SOME DISPERSION RULES

1. The delay between two states of polarizations should be held to 0.08 to 0.25 T, where T is the bit period.

2. Alternatively, acceptable dispersion ($\Delta\tau$) may be defined as[2]

$$\Delta\tau < T/4$$

3. The information bit rate is expressed as

$$R_b < 1/(4\Delta\tau)$$

Discussion

Most media (e.g., the glass of a fiber's core) have a slightly different index of refraction for the different states of polarization; therefore, the velocity of light in the fiber is slightly different for these different states. This is called *birefringence* and is seen in high school physics experiments with calcite. When light travels many kilometers through a fiber, the difference results in the pulse noticeably increasing in length and thus time duration. Remember, a typical pulse lasts only a few tens to hundreds of picoseconds, so the $\Delta\tau$ can grow rather large through a typical fiber link. Also, when light enters a fiber, some rays enter almost straight down the axis, while others approach the acceptance angle. The rays that enter near the axis are reflected fewer times from the cladding and travel a straighter path. Thus, they arrive sooner than the other rays, which enter just at the acceptance angle and bounce from side to side. Thus, a pulse of light will have its intensity reduced and spread out as a result of these modal delays. The pulse-mode dispersion is a function of both of these phenomena.

The increase can be calculated from

$$\Delta T = \sqrt{\sum_{k=1}^{n} D_{PMD}(k)^2 L(k)}$$

where $T =$ bit period
$D_{PMD}(k) =$ polarization-mode dispersion (PMD) parameter of the kth fiber segment as a function of k, the propagation constant
$L(k) =$ segment's length

This can be calculated by the above equation for several fiber segments of a pulse's path, and Goldstein, Lih-Yuan, and Walker[1] suggest keeping the $\Delta\tau$ to less than 10 percent of the pulse's total time.

Andre Girard[2] points out that, if the percentage of tolerable broadening by the receiver is established at about 10 percent of the bit period, it corresponds to a power loss of 1 dB (power budget accounting for polarization-mode dispersion). However, the tolerable 10 percent is also based on statistical broadening of the pulse due to PMD, as the pulse does not broaden with a fixed value as in chromatic dispersion; it varies all the time. Therefore, 10 percent essentially corresponds to the "most-of-the-time broadening." If you look at a Maxwell distribution, that "most-of-the-time" corresponds to the average value of the distribution; that is, 50 percent of the values are on one side, and 50 percent are on the other side of the distribution. But remember that this distribution also has a maximum value at a certain low probability.

Incidentally, typical values for D_{PMD} are about 1 ps/$\sqrt{km}$ for typical embedded fiber but can be much lower for twenty-first-century fibers made to accommodate 40 Gb or more. Dispersion does vary by wavelength for these fibers within common telecom bands (so it may be different from one DWDM wavelength to another). Andre Girard[2] points out that, due to "the random contribution of mode coupling, the 'normalized' value is referenced to the square root of the length unit, resulting in units of ps/$\sqrt{km}$ or ps/$\sqrt{m}$."

Kartalopoulos[4] indicates that the "4" in the very first equation of this rule is a typical value for the dispersion factor, which is a transmission design factor, and this constant is almost never more than 5. If it is 3 or less, it is more acceptable.

References

1. E. Goldstein, L.-Y. Lin, and J. Walker, "Lightwave Micromachines for Optical Networks," *Optics And Photonics News* 12(3), p. 62, 2001.
2. A. Girard et al., *Guide to WDM Technology and Testing*, EXFO Electro-Optical Engineering Inc., Quebec City, Canada, pp. 87–89, 2000.
3. Private communications with A. Girard, 2002.
4. S. Kartalopoulos, *Introduction to DWDM Technology*, IEEE Press, Piscataway, NJ, pp. 46–47, 2000.

CORRELATION WAVELENGTH FOR DIFFERENTIAL GROUP DELAY

The correlation wavelength for differential group delay (DGD) is

$$\Delta\lambda_c = \frac{3.2}{(DGD)}$$

where $\Delta\lambda_c$ = correlation wavelength in nanometers
       DGD = mean differential group delay in picoseconds

Discussion

Noutsios[1] states that, frequently, "The measurement error is derived using the correlation wavelength, $\Delta\lambda_c$, which is a measure of the wavelength separation between independent instances of DGD." It can be shown to be inversely related and normalized as in the above equation if measured in nanometers and if the DGD is in picoseconds.

Single-mode fibers transmit two orthogonal states of polarizations at slightly different speeds. This difference in index of refraction as a result of polarization is the definition of *birefringence*. These different speeds translate into pulse spreading and are measured as a differential group delay or DGD. The relations between DGD and PMD are simply normalization to the square root of the fiber length (for long fibers); DGD thus manifests itself as PMD. In the rule "PMD and DGD Statistics" (p. 260), a more thorough and complex relationship between these two terms is provided. Like PMD, the total DGD of a link can be determined by root-sum squaring the individual sublink DGDs as

$$DGD_{total} = \sqrt{(DGD_1)^2 + (DGD_2)^2 + (DGD_3)^2 + (DGD...)^2}$$

References

1. P. Noutsios et al., "Experimental and Theoretical Investigations of High PMD Impairments on an OC-192 Field System," *Proceedings of the National Fiber Optic Engineers Conference*, 1999.
2. C. Poole, J. Winters, and J. Nagel, "Dynamical Equation for Polarization Dispersion," *Optics Letters* 16(6), pp. 372–374, 1991.

DISPERSION PENALTY

The penalty due to dispersion as a function of data rate can be calculated as follows if the fiber's total bandwidth is known:

$$P_d = 1.22\left[\frac{BR}{BW}\right]^2$$

where P_d = dispersion penalty (in dB)
   BR = bit rate in gigabits per second
   BW = bandwidth in gigahertz

Discussion

Sometimes, only the total bandwidth in megahertz of the fiber is known. If one is sure that this is the total allowable bandwidth based on both modal as well as chromatic dispersion, then the above equation can be used to estimate the dispersion penalty in decibels.

To quote the reference,

> For typical communications grade fiber, the dispersion penalty for a 20 km link is about 0.5 dB. Dispersion is usually minimized at wavelengths near 1310 nm; special types of fiber have been developed which manipulate the index profile across the core to achieve minimal dispersion near 1550 nm, which is also the wavelength region of minimal transmission loss.[1]

This rule is closely related to other rules in this chapter. In fact, in a way, it is even just a restatement (in different form) of another rule. Note that the familiar dispersion loss is proportional to the square of the data rate.

Be careful of the definition of bandwidth given above, as very old legacy fibers frequently give this figure in megahertz rather than gigahertz.

Incidentally, it has been decided at a 2002 ITU-T meeting that, for 40 Gb/s transmissions, a 3-dB power penalty would be allowed for taking into account dispersion, 2 dB for chromatic dispersion, and 1 dB for PMD.

Reference

1. C. DeCusatis and G. Li, "Fiber Optic Communication Links (Telecom, Datacom and Analog)," in *Handbook of Optics*, Vol. 4, M. Bass, Ed. in Chief, McGraw-Hill, New York, pp. 6.9–6.14, 2001.

Fiber Dispersion per Unit Length Estimates

1. For graded index fibers, the lowest dispersion per unit length is approximated by

$$\frac{\delta\tau}{L} \approx \frac{n_1 \Delta^2}{10\,c}$$

where $\delta\tau$ = pulse width, full width at the half-power point
L = length of the fiber
n_1 = core index
c = free-space velocity of light
Δ = maximum relative index between core and cladding (see p. 227)

2. For a step-index fiber, the dispersion per unit length is

$$\frac{\delta\tau}{L} = \frac{n_1 \Delta}{c}$$

3. Additionally, for step-index single-mode fibers, waveguide dispersion is approximated by

$$D_{wg} \approx -\frac{0.025\lambda}{a^2\,c n_2}$$

where λ = free-space wavelength
a = core radius
n_2 = index of the cladding

Discussion

The above equations allow the user to estimate the dispersion from a typical fiber based on its physical properties. Obviously, these rules do not accommodate specially doped or special-material fibers designed for special dispersion characteristics.

The difference between the first and second equation reflects the hard boundary in the core to cladding index of refraction change. This is attributed to Δ, the index between the core and cladding, and is generally 0.01.

The reference states,

Bandwidth is inversely proportional to dispersion, with the proportionality constant dependent on pulse shape and how bandwidth is defined. If the dispersed pulse is approximated by a Gaussian pulse with δ_t being the full width at the half-power point, then the −3-dB bandwidth B is given by[1]

$$B = \frac{0.44}{\delta\tau}$$

Reference

1. I. Jacobs, "Optical Fiber Communication Technology and System Overview," in *Handbook of Optics*, Vol. 4, M. Bass, Ed. in Chief, McGraw-Hill, New York, pp. 2.2–2.3, 2001.

EFFECTIVE INTERACTION LENGTH FOR NONLINEAR EFFECTS

One can assume that the maximum length of a fiber transmission, when limited by nonlinear effects, can be roughly summarized as

$$L_{eff} = \frac{1 - e^{-\alpha L}}{\alpha}$$

where L_{eff} = effective fiber length for nonlinear interaction
α = attenuation coefficient
L = fiber link length

Discussion

Most nonlinear effects depend on intensity, so they fall off as the intensity decreases as a wave propagates down a fiber. This is analogous to Beer's law for atmospheric transmission; the intensity falls off as the exponential of the path length.

As the reference states, "Nonlinear effects tend to disappear as the wave progresses down the fiber. This results in a limited effective length over which the fiber nonlinearity influences the wave."[1]

The above applies only to unamplified links. When optical amplifiers are applied to the link, each amplified link must be recalculated for its nonlinear effects. For long fibers, the L_{eff} is on the order of 25 km in the C band as L_{eff} becomes approximately $1/\alpha$.

Reference

1. T. Stern and K. Bala, *Multiwavelength Optical Networks*, Prentice Hall, Upper Saddle River, NJ, p. 189, 2000.

MAXIMUM BIT RATE VS. FIBER LENGTH DUE TO DISPERSION

1. One can assume that the maximum length of a fiber transmission when limited by dispersion is approximately[1]

$$R_{max} = \frac{1}{\lambda}\sqrt{\frac{0.25\,c}{DL}}$$

where R_{max} = maximum bit rate
c = speed of light
D = fiber dispersion
L = fiber link length
λ = wavelength

2. Alternatively, Hobbs[2] gives the following simple equation, allowing one to estimate the link length when limited by intermodal dispersion:[2]

$$L \times BW < \frac{cn_1}{\pi(NA)^2}$$

where L = link length
BW = bandwidth
n_1 = core index
NA = numerical aperture

3. Additionally, Kapron and Dori[3–6] state the maximum transmission length (L) in kilometers as limited by PMD as the simple but elegant

L (km)= $10^4/[$(bit rate in Gb/s) (PMD coefficient in ps/$\sqrt{km}$)$]^2$

4. Judy[7] rewrites the above as

$$(B^2)(L)(D) = 100000 \ [(Gb/s)^2 \ ps/nm]$$

where B = bit rate
L = system length
D = fiber dispersion

5. Lastly, Zhao and Asundi[8] give the following relationship (based on the work of P. S. Henry):

$$B \leq \frac{1}{4DL\sigma_\lambda}$$

where σ_λ = rms spectral width of the signal

Discussion

Dispersion stretches a laser pulse because the different frequencies (chromatic dispersion) or polarizations (polarization mode dispersion) or different modes (modal dispersion) experience a different index of refraction and thus travel down the fiber with different velocities. Polarization mode dispersion stretches a laser pulse because the different polarizations experience a different index of refraction and thus travel down the fiber at slightly different speeds. This effect can easily stretch a pulse into a neighboring pulse.

Chromatic dispersion is less statistical and more deterministic. The bit-rate cannot exceed a value R_{max}, where $R_{max} = k/(DL\sigma)$, where σ is the pulse spectral width. Stern and Bala[1] point out that a reasonable value of k is 0.25, indicating a pulse overlap into the next bit interval of approximately 25 percent. From the basic relationship of wavelength and frequency, we have $\sigma = \lambda^2/cT$, and T is the pulse width in time. Assuming that R is approximately $1/T$, the first of the above equations will result.

To describe the situation another way, different order modes have different phase and group velocities. This limits the length of a step-index fiber as well as a gradient-index fiber. As Hobbs[2] points out, "We can feel pretty safe as long as the phase spread between modes is below a radian or so at the highest modulation frequency, but we expect to be in big trouble when it gets to half of a cycle." This leads to the second equation. The term on the right of the second equation is based on ray tracing of a ray that is bouncing off of the core-cladding boundary at the critical angle.

Polarization-mode dispersion (PMD) is a statistical phenomenon, and it depends on the root of the fiber's link length as indicated above. The mean PMD should not exceed one-tenth the bit time period. If one assumes a high level of mode mixing and ignores the nonfiber contributions to PMD, then the third equation can be written.[3–6]

References

1. T. Stern and K. Bala, *Multiwavelength Optical Networks*, Prentice Hall, Upper Saddle River, NJ, p. 185, 2000.
2. P. Hobbs, *Building Electro-Optical Systems, Making It All Work*, John Wiley & Sons, New York, p. 248, 2000.
3. F. Kapron et al., "Polarization-Mode Dispersion: Should You be Concerned?" *Proceedings of the National Fiber Optic Engineers Conference*, p. 675, 2001.
4. A. Dori et al., "Field Observation of Polarization Mode Dispersion Induced Impairments to 10 GB/s Transmission in Jackson, Mississippi," *Proceedings of the National Fiber Optic Engineers Conference*, 2001.
5. P. Noutsios, "PMD Assessment of Installed Fiber Plant for 40 Gbit/s Transmission," *Proceedings of the National Fiber Optic Engineers Conference*, p. 1343, 2001.
6. F. Kapron, "System Considerations for Polarization Mode Dispersion," *Proceedings of the National Fiber Optic Engineers Conference*, 1997

7. A. Judy, "Dispersion Compensation for Future High Bit Rate Systems," *Proceedings of the National Fiber Optic Engineers Conference*, 2001.
8. B. Zhao and A. Asundi, "Strain Microscope with Grating Diffraction Method," *Optical Engineering*, 38(1), pp. 170–180, Jan. 1999.

PMD and DGD Statistics

1. Polarization-mode dispersion (PMD) measurement uncertainty is

$$\frac{0.66\lambda}{\sqrt{DGD\ \Delta\lambda}} \tag{1}$$

where λ = wavelength in microns
   DGD = differential group delay (DGD) in picoseconds
   $\Delta\lambda$ = wavelength span for the measurement in nanometers

and

$$\text{RMS (DGD)} = \sqrt{\frac{3\pi}{8}} \approx 1.0854 \times \text{mean (DGD)} \tag{2}$$

2. The magnitude of the PMD experiences a Maxwellian distribution such that

$$p(I) = \frac{32I^2}{\pi^2 M^3} \exp\left(-\frac{4I^2}{\pi M^2}\right) \tag{3}$$

where $p(I)$ = probability density function of PMD
   I = instantaneous PMD
   M = mean PMD

3. The rms of the PMD is

$$M\sqrt{\frac{3\pi}{8} - 1} \approx 0.422\,M \tag{4}$$

Discussion

Both PMD and DGD are highly related and manifestations of the same phenomena. PMD as a definition is the root mean squared value of the DGD with the inclusion of the root fiber length. This phenomenon results from the slightly different values for the index of refraction for different polarizations and, as a practical manner, is statistical in nature. It is statistical because it varies from fiber to fiber, and there are even slight but real variations from one segment of a fiber to another segment of a fiber, even from one meter to the following meter of fiber. This difference is caused by the combination of manufacturing tolerances, temperature, vibration, and stress.

Equation (1) gives the DGD uncertainty as a function of the measurement's optical bandwidth.

Equation (2) really provides the distinction between the mean and rms of a DGD measurement or specification. This equation underscores the importance of knowing whether a given DGD is the mean, rms, or average. They are all related statistically but can be substantially different and must be applied appropriately in systems analysis.

Equation (4) above provides a handy relationship between the rms and mean pulse-mode dispersion. Again, be sure that you know whether the rms, mean, peak, or other definition of PMD is being specified.

Kapron[1] also points out that a powerfully practical concept is the complement of the Maxwellian distribution, which gives the probability that the PMD lies above a given value. Moreover, this can be multiplied by the number of minutes per year to yield a practical figure of merit for systems.

References

1. F. Kapron, "System Considerations for Polarization-Mode Dispersion," *Proceedings of the National Fiber Optic Engineers Conference*, 1997
2. B. Curti et al., "Statistical Treatment of the Evolution of the Principal States of Polarization in Single-Mode Fibers," *IEEE/OSA Journal of Lightwave Technology*, 8, pp. 1162–1166, Aug. 1990.
3. N. Bergano, C. Poole, and R. Wagner, "Investigation of Polarization Dispersion in Long Lengths of Single-Mode Fiber Using Multilongitudinal Mode Lasers," *IEEE/OSA Journal of Lightwave Technology* 15, pp. 1618–1190, Nov. 1987.
4. E. Murphy et al., "Measurement of Dispersion-Compensating Module Polarization-Mode Dispersion Statistics," *Proceedings of the National Fiber Optic Engineers Conference*, 1997.

PROPOSED PMD COEFFICIENTS

Bit rate (Gb/s)	Ten percent of the time between pulses (in ps), average DGD	PMD coefficient for a 400-km fiber (ps/√km)	Maximum fiber length for STM–64 OC–192 spec. of 0.5 ps/√km (km)
2.5	40	≤2	6400
10	10	≤0.5	400
20	5	≤0.25	100
40	2.5	≤0.125	25
60	1.67	≤0.083	11
80	1.25	≤0.0625	6.25
100	1	≤0.05	4

Discussion

A fiber's polarization-mode dispersion (PMD) is a critical consideration and frequently is the limiter for the length of fiber over which the data can be transmitted for high-data-rate transmissions (but usually is of far less concern for OC-48 and below). If one assumes a 99.994 percent probability that the power penalty will be less than 1 dB for 10 percent of the bit period, then the polarization-mode dispersion must satisfy the restrictions in the table.

One can clearly see that increasing modulation beyond 10 Gb is not practical with legacy cable with up to a 17 ps/√km PMD coefficient of dispersion. Moreover, rates exceeding 40 Gb/s are not practical until break-throughs in technology are implemented, and even existing OC-192 fibers can rarely support 40 Gb/s. Moreover, it is not practical for a network to place repeaters every 4 km for a 100 Gb/s modulation rate. This is one of the reasons for sending more lasers (or a different frequency) down a given fiber to increase transmission (DWDM), albeit at lower data rates per wavelength.

Differential group delay (DGD) occurs when two identical light pulses have closely spaced optical frequencies and propagation delays, and it is addressed by other rules in this chapter.

At the time of this writing, only the 10 Gb/s PMD rule of 0.5 ps/√km has been approved by international standards organizations such as ITU-T. The 40 Gb/s specifications is under discussion but, so far, no consensus has been reached.

Reference

1. Private communications with A. Girard, 2002.

PULSE SPREADING PER KILOMETER

The bandwidth-distance product for an acceptable data rate is usually between 2.0 and 2.0 GHz-km.

Discussion

Based on the technologies of the 1980s, Li[1] gives a figure of merit for using fibers in communications as the product of the bandwidth and the distance that it can carry a signal before pulse spreading leads to an unacceptable error rate. The value is usually between 2.0 and 2.4 GHz-km.

This is a simple rule, based on empirical observations of current multimode fibers. Thus, it depends greatly on the state of the art and will increase in the future. It does not account for absorption losses in the fiber or couplers and does not apply to specially configured fibers with tailored dispersion coefficients.

This rule can be used to estimate the distance a fiber can carry a given data rate signal or data rate.

The transmission bandwidth of a fiber is frequently driven by dispersion of the various modes. This arises because different modes have different group velocities. Dispersion also occurs in transmitting optical elements because of the variation of the index of refraction as a function of the wavelength of the light. Li[1] indicates that the modal dispersion can be minimized by a "near-parabolic refractive index profile in the fiber cross-section." The greater the length of the transmission, the more pronounced this effect becomes.

Additionally, for step-index fibers, travel time variation between the axial mode and the higher-order modes of a cladded fiber cable can cause undesirable pulse stretching, which may limit the data rate of the fiber. According to Hecht,[2] the time difference between the axial (shortest time) mode and the slowest mode is Δt, which can be expressed as

$$\Delta t = \frac{Ln_f}{c}\left(\frac{n_f}{n_c} - 1\right)$$

where Δt = time difference between modes
L = length of the fiber
n_f = index of refraction of the fiber
c = speed of light
n_c = index of the fiber cladding

If one wishes to transfer 500 Mb of information over 50 km, one can divide 2.5 GHz-km by 50 and estimate that the maximum data rate is 0.05 Gb (or 50 Mb) per second. Thus, the information can be sent over 50 km in only 10 s. However, one should also check the Δt to verify the maximum data rate.

Pulses injected into the cable cannot exceed one pulse every $2\Delta t$ so that the pulses can be discerned at the receiver. A 50-km cable with a Δt of 10 ns can accept pulses at a rate of one per 20 ns, which is equivalent to 50 million pulses per second.

This rule is closely related to other rules in this chapter. In fact, in a sense, it is even a restatement (in different form) of another rule. Note that the familiar dispersion loss is proportional to the square of the data rate.

This rule might only apply to intermodal dispersion (in MMF) and not intramodal dispersion.

References

1. T. Li, "Lightwave Telecommunication," *Physics Today*, May 1985.
2. E. Hecht, *Optics*, Addison-Wesley, New York, p. 175, 1990.

POLARIZATION-MODE DISPERSION ROOT SUM SQUARED FOR NETWORK

The total PMD for cascaded links is the *root sum squared (RSS)* of the PMD of all of the links, or

$$PMD_t = [(PMD_1)^2 + (PMD_2)^2 + (PMD_3)^2 + (PMD_n)^2]^{1/2}$$

where PMD_t = total PMD that a communications packet experiences
PMD_1 = PMD of the first link
PMD_n = PMD of the nth link

Discussion

Polarization-mode dispersion (PMD) can often limit the length and data rate of transmission for high-data-rate links. The PMD that will be experienced by a packet traveling though multiple cascaded links is root-sum-squared.

Girard states,

> . . . nine out of ten spans in a network each have a PMD of 0.2 ps, but the tenth one has a PMD of 2 ps, the total PMD would be 2.008 ps. In other words, the single bad section dominates the overall figure. Thus, all spans in the network must be tested; it cannot be assumed that if a few spans show low PMD that, overall, they will be acceptable.[1]

Girard also reiterates the age-old advice that the PMD measurement should be taken over environmental and operational conditions that will be experienced by the actual network. DGD will "Maxwellize" over wavelength, over time, over temperature, and over everything if you have sufficient statistical behavior of the variable. It means, for instance, that if you keep everything else constant and wait sufficiently long, the DGD behavior will finally follow a Maxwellian distribution. This is very difficult when you deal with external environment, and it is almost impossible to keep the environment constant. So if you go in the field and make a PMD measurement, you must make it very quickly over as broad a range of wavelengths as possible. This is why the interferometric method has been approved for PMD field test. But even if you make the test fast enough, the environment will change and, consequently, it is important to assess that change by making repetitive measurement over different environmental conditions—for instance, over a day or over a week, and over varying temperature and environmental weather conditions—to evaluate these effects to PMD.

In another rule ("Noise Root Sum of Squares," Chap. 8), a handy method for quickly calculating a RSS is given, and it deserves to be repeated here. A rough answer can be obtained by just taking the largest value. If the separate components are roughly the same, then multiply the largest value by 1.4 for the combined figure. It often works for larger variance as well. Here's an example that proves the point. Suppose the system has four terms of 5, 4, 2, and 1. Using the quick-approximation method, we would estimate the error as $1.4 \times 5 = 7$. Doing the root sum of squares, we get 6.8.

Reference

1. A. Girard et al., *Guide to WDM Technology and Testing*, EXFO Electro-Optical Engineering Inc., Quebec City, Canada, pp. 45–47 and 93–94, and private communications with A. Girard, 2002.

SECOND-ORDER POLARIZATION-MODE DISPERSION

1. The second-order polarization-mode dispersion (PMD_2) can be estimated from[1]

$$PMD_2 = \frac{2\pi c PMD_1^2}{1.7\lambda^2}$$

where PMD_2 = second-order polarization-mode dispersion
  c = speed of light
  PMD_1 = first-order polarization-mode dispersion
  λ = wavelength

2. Kapron[2] gives

$$PMDC_2(ps/nm\text{-}km) \approx \left[\frac{1043}{\lambda(nm)}\right]^2 \times PMDC_1^2(ps^2/km)$$

where $PMDC_2$ = second-order polarization-mode dispersion
  $PMDC_1$ = first-order polarization-mode dispersion based on rms DGD

3. Girard[5] gives a most simple and potentially useful rule relating PMD_1 to PMD_2

$$PMD_2 = \frac{PMD_1^2}{1.7}$$

Discussion

It is well known that polarization-mode dispersion (PMD) can often limit the length and data rate of transmissions, as elaborated in other rules collected in this chapter. Unfortunately, there are second, third, and other orders of PMD that can also develop and, in stressing applications, can cause problems and issues. There are orders as in chromatic dispersion (CD) if you properly define the expansion of beta, beta 1 (the first-order derivative of beta, which is the index of refraction times the propagation constant k) being the group delay, beta 2 the second-order derivative being the dispersion, and so on. Second-order PMD is the same. First-order PMD is related to the first-order derivative of birefringence, meaning first-order derivative of (beta y – beta x), second-order being the second-order derivative of the birefringence, and so forth. Therefore, you can have third-order PMD and so on.[5] The first-order PMD scales as the root of the fiber length, while the second-order scales linearly with the fiber length.

The second-order PMD is the change of PMD with wavelength and is related to the first-order phenomenon. As pointed out by the Girard,[1] "a first-order PMD coefficient of 0.5 ps/$\sqrt{km}$ will be accompanied by a second-order effect of about 0.15 ps/nm-km." The second-order effect is not necessary negligible and should be considered in high-speed optical telecommunication transmission errors, especially as the first-order is decreased thorough various techniques.

Kapron states, "There are two wavelength variations that contribute to the second-order PMD: variation of the DGD and variation of the rotation of the PSPs."[2] (In the Poincaré sphere representation, this corresponds to a change in both the magnitude and direction of the polarization vector 22, and the latter usually dominates 23.) A DGD vs. wavelength measurement alone is not sufficient for characterizing second-order PMD. In the limit of full-mode mixing, the second-order coefficient is related to the first-order PMD coefficient of

$$PMDC_2 = \frac{2\pi c}{\lambda^2 \sqrt{3}} \times PMDC^2$$

The second rule given above is a practical simplification of the above equation. Kapron goes on to point out, "For a cable with a first-order PMD coefficient of 0.1 ps/$\sqrt{km}$, the second-order PMD coefficient at 1310 nm is 0.0063 ps/km-nm. However, this combines with chromatic dispersion and over long lengths can be significant."

Moreover, Cyr[3] points out that any system can be adversely affected by PMD_2 if it meets any of the following criteria:

1. The CD coefficient is greater than 2 ps/nm-km

2. Bit rates of >10 Gb/s for digital systems

3. Link lengths of greater than 100 km

4. A fiber with a PMD coefficient of >0.05 ps/$\sqrt{km}$

The last equation from Girard assumes random mode coupling and an infinite ratio of the fiber length to the coupling length (L/h infinite), and it pertains only to the average value over wavelengths and does not provide any spectral details.

Incidentally, the "1.7" in the above equations comes from the square root of 3.

Second-order PMD will play a key role at data rates above 40 Gb/s when PMD compensation is used. Second-order PMD will provide a polarization-dependent chromatic dispersion (PD group delay) that will give an non-precise compensation and residual noise to the pulse. Moreover, not only PMD_1, PMD_2, and CD will play a role, but there are also phase-related nonlinear effects such as self-phase modulation, cross-phase modulation (in DWDM systems), and techniques such as pre-chirping and forward error correction (FEC).[4]

References

1. A. Girard et al., *Guide to WDM Technology and Testing*, EXFO Electro-Optical Engineering Inc., Quebec City, Canada, pp. 87–90, 2000.
2. F. Kapron, Systems Considerations for Polarization-Mode Dispersion, *Proceedings of the National Fiber Optic Engineers Conference*, 1997.
3. N. Cyr, M. Breton, and G. Schinn, "Second Order PMD: What Is It and How Can It Affect Your System?" *Proceedings of the National Fiber Optic Engineers Conference*, 1997.
4. Private communications with A. Girard, 2002.

SOLITON CONDITIONS

Soliton conditions are met when the nonlinear and dispersion lengths are the same.

Discussion

Dispersion length is covered elsewhere and is described by the following equation:

$$z_d = \frac{2\pi c 0.322 \tau^2}{\lambda^2 D}$$

where τ = pulse width
  D = dispersion

The nonlinear length is defined as the distance the pulse travels before its spectral width doubles,

$$z_{nL} = \left(\frac{2\pi n_2 I_0}{\lambda} \right)^{-1}$$

Here, I_0 is the initial intensity of the pulse. Combining the two equations allows us to compute a relationship between soliton peak power, wavelength, fiber index, dispersion, and pulse width.

$$P_0 = \frac{\lambda^3 D A}{0.322(4\pi^2) c n_2 \tau^2}$$

where n_2 is the nonlinear refractive index. That is, n_2, when multiplied by the intensity in the fiber, defines a small change to the zero intensity index of refraction. For silica fibers, n_2 has a value of about 3×10^{-16} cm^2/W. Note that, since intensity has the units of W/cm^2, the index is unitless, as expected.

Reference

1. P. Mamyshev, "Solitons in Optical Fiber Communication Systems," Chap. 7 in *Fiber Optics Handbook*, M. Bass, Ed. in Chief, and E. Van Stryland, Assoc. Ed., McGraw-Hill, New York, p. 7.4, 2001.

DGD Soliton Pulse Broadening

$$\tau(z) = \sqrt{1 + \frac{0.1081 DGD^2}{\tau_0^2}}$$

where $\tau(z)$ = rms (root mean square) width of the pulse in picoseconds
   τ_0 = pulse width
   DGD = mean differential group delay in picoseconds

Discussion

Solitons are being reconsidered for data transmission in the region of 100 Gb/s and for some ultra-long-haul applications. A soliton pulse consists of a wave group in which the polarizations are quantum mechanically tied to one another. As such, solitons exhibit a "self-trapping" of the polarization states and resist the classic pulse broadening caused by the birefringence in the fiber material. However, as Xie[1] explains, a soliton is not immune to effective PMD pulse spreading, because "the soliton pulse propagating in the fibers with PMD is continuously perturbed by the randomly varying birefringence and will radiate and lose energy, thereby also broadening during propagation, although at a lower rate than a linear pulse. Thus the variations in the fiber's birefringence result in energy to be depleted from the soliton to remain self-trapped."

Xie derives the above rule from analyzing the energy loss as the soliton travels down a fiber. With some manipulation and elegant math, Xie derives the following relationship for the energy loss by assuming the PMD acts as a radiative wave taking energy from the soliton.

The derivation of the above rule starts with

$$\frac{dE}{d\xi} = -\frac{32}{729} \frac{E^3 dt^2}{\xi_c}$$

where $\dfrac{dE}{d\xi}$ = energy loss rate

By solving this equation and using real units, the following can be written:

$$\frac{E(z)}{E(0)} = \frac{1}{\sqrt{1 + \pi D_p^2 z/(24 T_0^2)}}$$

where T_0 = pulse width
   D_p = PMD coefficient

With some refinement, the above can be rewritten as the simpler equation given in the rule.

Andre Girard[2] points out some important issues with solitons in general. Solitons are very narrow pulses with a long tail (at least those with sech and sech^2 shapes). They are very sensitive to noise such as ASE from optical amplifiers. That is why Raman amplifiers, with their very low noise levels, may potentially enable future use of soliton transmissions. Another problem due to the tail is the overlap risk with other adjacent channels in DWDM systems.

Also, a very narrow pulse brings a very wide spectrum that limits the efficiency of the transmissions. Immunity to noise could be achieved by careful dispersion mapping and noise control at each amplifier site if the link design would allow that.

References

1. C. Xie et al., "Effectiveness of PMD Mitigation Using Solitons," *Proceedings of the National Fiber Optic Engineers Conference*, pp. 656–663, 2001.
2. Private communications with A. Girard, 2002.

TIMING JITTER

Timing jitter can be caused by acoustic interaction of pulses.

Discussion

This is a result of the electrostrictive effect in which each pulse in the fiber sets up an acoustic wave. These waves modulate the index of refraction of the fiber, causing variation in the arrival time of what are initially equally spaced pulses. The timing jitter σ is

$$\sigma = 4.3\frac{D^2}{\tau}\sqrt{(R-0.99)}L^2$$

where L = fiber length in megameters
        τ = pulse width

Thus, undamped cables that carry high data rates, laid in highly acoustic environments (e.g., a quarry, along a highway, or though a venue that has frequent rock concerts), can have an occasional timing jitter issue. This effect may become analogous to that of dispersion. No one really cared about PMD when data rates were less than a gigabit per second. However, it became critical as rates approached 40 Gb/s. As data rates approach 100 Gb/s, the electrostrictive properties of fibers will become more important.

Reference

1. P. Mamyshev, "Solitons in Optical Fiber Communication Systems," Chap. 7 in *Fiber Optics Handbook*, M. Bass, Ed. in Chief, and E. Van Stryland, Assoc. Ed., McGraw-Hill, New York, p. 7.7, 2001.

TRANSMISSION DISTANCE CAN BE LIMITED BY CHROMATIC DISPERSION

Transmission distance (L_D) can be limited by chromatic dispersion and the squared inverse of the bandwidth (B).

Discussion

As mentioned in the previous rule, the equation that describes the distance limit is

$$L_D = \frac{1}{BD\Delta\lambda}$$

Note that the rule is correct, since the spectral bandwidth ($\Delta\lambda$) is proportional to B, hence the inverse square behavior. Beyond this distance, the pulse broadening exceeds the size of the pulses. A typical example has $B = 2.5$ Gb/s, dispersion (D) 14.68 ps/nm/km and bandwidth of 0.025 nm, limiting the system to 1090 km. An increase of bandwidth to 10Gb/s limits the range to 60 to 70 km. The selection of 14.68 ps/nm/km of course depends on the wavelength of operation of the system. A typical fiber might have a wavelength of zero dispersion around 1300 nm with a slope of 0.075 ps/nm^2/km. If we operate the system at 1550 nm, we get a chromatic dispersion of 14.68 ps/nm/km.

Reference

1. A. Willner and Y. Xie, "Wavelength Domain Multiplexed (WDM) Fiber-Optic Communication Networks," Chap. 13 in *Fiber Optics Handbook*, M. Bass, Ed. in Chief, and E. Van Stryland, Assoc. Ed., McGraw-Hill, New York, p. 13.5, 2001.

PROPAGATION CONSTANT IS A FUNCTION OF WAVELENGTH

Waveguide dispersion results from the fact that the propagation constant of a fiber is a function of wavelength ($2\pi/\lambda$).

Discussion

For a step-index single-mode fiber, waveguide dispersion is approximated by

$$D_{wg} \approx -\frac{0.025\lambda}{a^2 cn_2}$$

where a = the fiber radius
        n_2 = index of the cladding

For a single-mode fiber (SMF), waveguide dispersion is small (about –5 ps/nm/km).

Reference

1. I. Jacobs, "Optical Fiber Communication Technology and System Overview," Chap. 2 in *Fiber Optics Handbook*, M. Bass, Ed. in Chief, and E. Van Stryland, Assoc. Ed., McGraw-Hill, New York, p. 2.4, 2001.

MAXIMUM DIFFERENTIAL GROUP DELAY IS THREE TIMES THE MEAN

It is generally accepted to consider that the maximum differential group delay (DGD) is three times the mean.

Discussion

Obviously, this must be understood in terms of a certain probability, as "maximum DGD" does not mean anything if you do not correlate it with a probability. For this three-times case, a probability of 4.2×10^{-5} is usually assumed.

Also remember that the network operator is going to have to accept this and the associated failures. It is certain that a problem will arise. He does not know exactly when, so he will have to plan for a certain level of quality of service and then accept the obligation to pay a penalty when the maximum DGD happens.

Reference

1. Private communications with A. Girard, 2002.

CHROMATIC VS. ANOMALOUS DISPERSION

In chromatic dispersion, where $D < 0$, the high frequencies (shorter wavelengths) travel with slower velocities. In anomalous dispersion $(D > 0)$, the higher frequencies (shorter wavelengths) travel faster.

Discussion

The above points out that, generally, shorter wavelength light travels somewhat more slowly than longer wavelengths. Thus, the effective index of refraction is larger for smaller wavelengths for typical fibers. However, the converse can exist as well.

DISPERSION LENGTH AND PULSE WIDTH

Dispersion length depends on the square of the pulse width.

Discussion

Dispersion length is defined as the distance a pulse travels before is broadened by a factor of $\sqrt{2}$ in time. An approximation for this distance (z_d) is

$$z_d = \frac{2\pi c 0.322\tau^2}{\lambda^2 D}$$

where D = dispersion
         τ = pulse width
         λ = wavelength

Reference

1. P. Mamyshev, "Solitons in Optical Fiber Communication Systems," Chap. 7 in *Fiber Optics Handbook*, M. Bass, Ed. in Chief, and E. Van Stryland, Assoc. Ed., McGraw-Hill, New York, p. 7.2, 2001.

UNEXPECTED GAIN FROM STIMULATED BRILLOUIN SCATTERING

Stimulated Brillouin scattering (SBS) can provide gain in a fiber.

Discussion

Although generally considered a deleterious effect, SBS can provide a beneficial gain. Although less useful than stimulated Raman scattering, SBS can be a source of gain for special applications. Values around 5×10^{-11} m/W have been reported but are useful only in very narrow frequency bands.

Reference

1. T. Brown, "Optical Fibers and Fiber-Optic Communication," Chap. 1 in *Fiber Optics Handbook*, M. Bass, Ed. in Chief, and E. Van Stryland, Assoc. Ed., McGraw-Hill, New York, p. 1.38, 2001.

11

Optics

It is easy to argue that the general field of optics has enabled all of the successes of the fiber optics industry. Indeed, from a historical point of view, it is plain to see that the theory of guided waves of optical-wavelength signals has existed since Maxwell's time. The real breakthroughs came when new materials were introduced that made long distance propagation a reality. That development, along with high-performance lasers and erbium-doped fiber amplifiers, has introduced a whole new world of capability.

The rules in this chapter are intended to introduce some basic optics concepts for the designer who is new to optical systems. They may be most useful for engineers who are familiar with designing telephone and other "copper-based" systems that do not involve optics. While we were not able, in this small space, to cover all of the key topics that should be known to anyone working in fibers or optics in general, we have included some important topics.

Perhaps the most important is a rather large rule that describes the often-overlooked topic of etendue[*] (typically 10^{-8} to 10^{-5} cm^2-sr). Similarly, we provide some rules related to the properties of Gaussian beams, a field that is always useful when dealing with light emitted from lasers. Almost everyone encounters a situation now and then in which a quick model of the shape of a Gaussian beam is useful, so we have included one. Additional rules about Gaussian beams can be found in Chap. 5, "Free-Space Optical Communications," and Chap. 13, "Laser Transmitters." We have included a rule that quickly describes the properties of diffraction in optical systems.

---

[*]Etendue (literally, *extent*) refers to the ability of an optical system to accept light. As a function of the source area and the solid angle into which it propagates, etendue is a limiting function of system throughput.

Finally, in the category of basic understanding, we include a rule that shows how one converts from optical frequency to wavelength and back. This is a particularly useful tool when one encounters the various experts in the field, some of whom work in either frequency or wavelength, for historical reasons. The typical application of this conversion process is to determine how the bandwidth of filters (sometimes expressed in gigahertz) converts to wavelength bands.

We also include some rules related to the Faraday effect and other polarization effects. Antireflection coating ideas are included to make sure that those intending to use or design coatings are equipped to do so.

We have included some practical rules related to keeping optics clean and aligned, and for making sure they are the right size for the intended application. Similarly, we have included some ideas on tolerancing optical systems, which is so critical for the manufacture and producibility of optical systems.

Perhaps the best news for the newcomer to the fiber field is that there are hundreds of books on optics that can augment what is included here.

ALIGNMENT PROBLEMS

Alignment is almost always better in the lab than in the field. The one exception seems to be when it is worse in the field than it is in the lab.

Discussion

This is based on empirical observations and applies to systems where no automated field adjustments are possible or included. A properly designed and toleranced optical system can develop slight alignment changes over the life of the system, which may not noticeably affect performance. With the advent of cheap computers and sensors, it is now possible to beat this rule by building a control system that senses errors and generates commands for their removal. For systems that do not include such correction features, one has to leave some margin for alignment in optics tolerancing and radiometric performance calculations. An optical design that only meets its performance requirements when precisely aligned will simply never meet its requirements in the real world.

Aligning a large train of optical elements is a black art. If the gods smile on you, you may achieve an alignment in your lab suitable for optimal performance. The performance is always made worse by misalignment. However, it is wise to allow for some degradation when the system is really put to use. Temperature gradients cause misalignment. Unstable optical and structural materials cause misalignment. These materials are usually the low-mass materials that are attractive for space applications. Thermal cycling causes misalignment. Dropping the optical trains on the ground tends to cause misalignment. Even the index of refraction of the air around the system varies, causing changes in focus. The list goes on and on . . . .

ANTIREFLECTION COATING INDEX

Antireflection coatings have no reflectance when their index of refraction is the square root of the index of refraction of the substrate and they are a quarter of a wavelength thick.

Discussion

That is, the thickness is $\lambda/4$ in the material, and $n_1^2 = n_o n_s$. Here, the subscript 1 refers to the index of refraction of the film material, o refers to the external medium (commonly, air which has an index of 1), and s refers to the index of the substrate on which the film is formed.

This rule works for normal incidence only and is based on Fresnel reflection coefficients (which can be derived from Maxwell's equations, if you must) and wavefront, interference theory, and basic ray tracing. For oblique incidence, division by the cosine of the incidence angle is required, as the path length of the light in the coating is longer.

Benefits are gained for antireflection coatings, even if the above is only partially met. This rule is useful in the design of coatings and underscores the nature of antireflection coatings.

Antireflection coatings should be applied to the surface of optics and focal planes. For high index of refraction materials (such as germanium) such coatings can almost double the throughput. If the above rule is met, the first surface reflection is reduced to a minimum at a given wavelength.

The reader should take note that the form of the reflectivity function when zero reflectance is not desired is as follows:

$$R = \frac{\left(n_o n_s - n_1^2\right)^2}{\left(n_o n_s + n_1^2\right)^2}$$

In this case, the thickness is still $\lambda/4$, but the indices are not selected for total nonreflection.

CLEANING OPTICS CAUTION

Dirty optics should be cleaned only after great deliberation and with great caution.

Discussion

This rule is based on empirical observations. Wiping off dirt often makes the surfaces more defective that the dirt did because

1. Most surfaces and all fingers have very fine abrasive dirt on them, which will scratch an optical surface
2. A few big areas of dirtiness are less harmful (scatter less light) than the myriad of long scratches left behind after removing the hunks, since scatter is a function of perimeter of the scratch.
3. Little particles can adhere very strongly (in proportion to their mass) and cannot be blown or washed off easily.
4. Washing mounted optics just moves the dirt into mounting crevices where it will stay out of reach, waiting for a chance to migrate back to where it is harmful.

Sometimes it is necessary to clean optics, especially if the contaminant is causing excessive scatter or if a transmitter will burn them into the coating. The shorter the wavelength, the more valid this rule becomes.

Reference

Private communications with W. Bloomquist, 1995.

THE "ETENDUE OR OPTICAL INVARIANT" RULE

At any plane in a lossless optical system, for a given wavelength, the product of the solid angle and the area of the ray bundles that the radiation is traveling through is a constant, or

$$C = A_o \Omega$$

where C = a numerical constant for a given detector pixel size and wavelength
A_o = useful area of the optics
Ω = solid angle field of view (be careful to include the effects of the index of refraction)

Additionally (with reference to Fig. 11.1 and the explanation provided in the "Discussion" section), we can write

$$A_d \Omega_i = A_s \Omega_o = A_o \Omega' \approx A_o \, (\text{IFOV}) \approx C_d \lambda^2$$

where A_d = area of the entire detector (or pixel in an array)
Ω_i = solid angle subtended by the detector in image space
A_s = area of interest in the scene
Ω_o = solid angle of the scene in object space
A_o = area of the optics
Ω' = solid angle of the detector projected into object space
IFOV = instantaneous filed of view of a detector (FPA) pixel
λ = wavelength (for a broad bandpass system, use the midpoint for this)
C_d = a constant determined by the pixel geometry's relationship to the blur diameter (see below); generally, for imaging systems, from 1.5 to 10

Discussion

The etendue relationship is a basic premise of radiometry and optics and is vitally important for systems employing optical fibers, as they have such

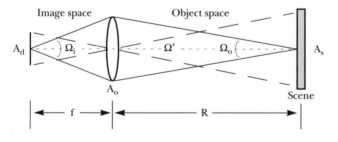

Figure 11.1

small etendue. It derives its foundation from the diffractive properties of optics. The law could be augmented by explicitly including the index of refraction, but we prefer to show it in its more conventional form and remember that the solid angle must be calculated with refractive properties in mind. When the index is included, as is necessary when discussing emergence of a beam from a fiber, etendue changes into $n^2 A\Omega$. Fibers usually have an etendue of 10^{-8} to 10^{-5} cm^2-sr.

In a diffraction-limited system, the blur diameter is equal to the Airy disk or 2.44 $(\lambda/D)f$, where D is the aperture diameter, and f is the focal length. If a square detector is matched to the blur, its area is the square of this, or $5.95(\lambda^2/D^2)f^2$. The solid angle seen by the detector is A_o/f^2, so

$$A_d\,\Omega_i \approx 5.95(\lambda^2/D^2)\,(f^2)\,(A_o/f^2) \approx 6\lambda^2$$

In systems that are not diffraction limited, the "6" (or C_d) is replaced by a larger number, but the important λ^2 dependence remains.

Similarly, a pixel (again, matched to the Airy disk) projected onto the scene has an area $[2.44(\lambda/D)R]^2$, where R is the range, and the solid angle is $[2.44(\lambda/D)R]^2/R^2$ or simply $5.95\lambda^2/D^2$ and (for an unobscured circular aperture),

$$A_o\Omega' \approx [5.95\lambda^2/D^2]\,[\pi D^2/4] \approx 4.7\lambda^2$$

The equations assume a lossless system; actual engineering calculations should include some transmission loss. It does not account for changes in index of refraction.

The generic etendue works regardless of aperture and pixel shape (it is frequently used by spectroscopists working with slits). Note that the premise for parts of the second equation assumes an unobscured circular aperture and a square pixel matched to a circular blur.

When properly applied, it allows one to estimate another system's (e.g., a competitor's) useful aperture, f/# or detector size. (If you don't know the speed of the system, assume an f/1; rarely are faster speeds achieved.) It provides a determination of collection aperture for radiometric applications.

- It provides a quick calculation of the wavelength, aperture size, or IFOV, if the other two are known.
- It can be used for estimates when coupling light into fibers, since there is a small cone, defined by the equation, that allows light acceptance into a fiber.

This rule goes by many different names. Spectroscopists like "etendue," ray tracers like "Lagrange theorem," and radiometry buffs like to say "optical invariant." It directly relates to other rules in this book.

The important relationship is that the useful aperture and the field of view are inversely related to each other. The numerical value of their actual relationship depends on the optical design, the chosen paraxial ray, and the height of the object or detector. The numerical constant isn't important (you can choose that based on your design or assumptions). What is important is the understanding that increasing one must result in the decrease of the other. Hence, large multiple-meter astronomical telescopes have small fields of view, and wide angle warning systems have very small apertures. You can have one but not the other . . . yup, life is tough. Additionally, a zoom lens will have a larger effective aperture when viewing its narrow field (resulting in a brighter image) as compared to its wide field.

As Longhurst puts it,

> In paraxial geometrical optical terms, the ability of an optical system to transmit energy is determined by a combination of the sizes of the field stop and the pupil in the same optical space; it is measured by the product of the area and the pupil in the same optical space; it is measured by the product of the area of one and the solid angle subtended at its center by the other. This is the three dimensional equivalent of the Helmholtz–Lagrange invariant or the Sine Relation.[6]

Given the size of a detector and the loose approximation that it can "sense" energy from one steradian of solid angle, the upper limits of the field of view and capture area in any associated optical system are immediately calculable.

References

1. Private communications with Dr. J. Kerr, 1995.
2. Private communications with Dr. G. Spencer, 1995.
3. R. Kingslake, *Optical Systems Design*, HBJ, Orlando, FL, pp. 36–38, 43–44, 1988.
4. A. Siegman, *Lasers*, University Science Books, Mill Valley, CA, p. 672, 1986.
5. C. Wyatt, *Radiometric System Design*, Macmillan, New York, pp. 36, 52, 1987.
6. R. Longhurst, *Geometrical and Physical Optics*, Longman, New York, pp. 465–467, 1976.
7. I. Taubkin et al., "Minimum Temperature Difference Detected by the Thermal Radiation Of Objects," *Infrared Physics and Technology* 35(5), p. 718, 1994.
8. P. Hobbs, *Building Electro-Optical Systems, Making It All Work*, Wiley Interscience, New York, p. 27, 2000.

GAUSSIAN APPROXIMATION TO DIFFRACTION

A Gaussian approximation to a diffraction light distribution can be found by matching the two curves at the $1/e$ point of the Gaussian. In that case, the appropriate value of σ of the Gaussian is

$$0.431 \; \lambda \; f/\# \text{ for a circular aperture}$$

$$0.358 \; \lambda \; f/\# \text{ for a square aperture}$$

where $f/\#$ = ratio of focal length to aperture

Discussion

The comparison of the Airy distribution of a circular aperture with a Gaussian provides this approximation. The computation of encircled energy or other characteristic of the diffracted field is quite complex. This approximation allows results to be obtained that are correct to within 2 to 10 percent.

This rule is quite useful in computations where integration of the blur spot of a diffraction-limited optic is required. This rule provides results that are good enough in most situations.

When the value of σ suggested above is used to approximate a diffraction spot, one gets

$$P(r) \; = \; \frac{1}{2\pi\sigma^2} \exp\left[\frac{-r^2}{2\sigma^2}\right]$$

where r = typical radial dimension

With this formulation, some calculations are considerably simplified as compared with trying to obtain the exact form derived from diffraction theory. For example, the energy distribution from a circular aperture is found to be a function made up of Bessel functions.

Reference

1. G. Cao and X. Yu, "Accuracy Analysis of a Hartmann-Shack Wavefront Sensor Operated with a Faint Object," *Optical Engineering* 33(7), p. 2331, July 1994.

GAUSSIAN BEAM RADIUS RELATIONSHIPS

$$r_{1/e^2} = 1/(2NA) \text{ or } \lambda/(\pi NA)$$

$$r_{99} = 1.57\, r_{1/e^2}$$

$$r_{3dB} = 0.693\, r_{1/e^2}$$

where r_{1/e^2} = radius (in microns) where the intensity is decreased by $1/e^2$ compared to its size at the beam waist

NA = numerical aperture

λ = wavelength

r_{99} = 99 percent of the power is included in a circle with this radius

r_{3dB} = 3-dB power density radius

Discussion

Laser spots passing through low-numerical-aperture optics tend to be Gaussian; thus, most laser beams in optical telecommunications can be represented as Gaussian beams.

The first equation, $r_{1/e^2} = 1/(2NA)$, is a simplification of the $\lambda/(\pi NA)$. The factor of 2 comes from π/λ, which is 2.02 for a wavelength of 1.557 microns.

Gaussian beams tend to be tighter than imaging spots with Airy disk patterns. Remember that the radius is in the same units as the wavelength, so if you use nanometers for the wavelength, the radius will be in nanometers.

Hobbs points out,

> The Gaussian beam is a paraxial animal: It is hard to make a good one of high NA. The extreme smoothness of the Gaussian beam makes it exquisitely sensitive to vignetting (which of course becomes inevitable as $\sin\theta$ approaches 1), and the slowly varying envelope approximation itself breaks down as the numerical aperture increases.[1]

Reference

1. P. Hobbs, *Building Electro-Optical Systems, Making It All Work,* John Wiley & Sons, New York, pp. 12–13, 2000.

HIGH NUMERICAL APERTURE SYSTEM DESIGN

The goal in designing a high numerical aperture (NA) lens system is to produce a system in which the lens aperture is nearly as large as the system focal length, while minimizing aberrations.

Discussion:

The numerical aperture is the cone angle through which an optical system collects or focuses the incoming light. It is critical for fiber optic applications. The NA is equal to the index of refraction multiplied by the sine of the acceptance angle. For an optical system with uniform illumination, the minimum focused spot size is

$$\text{Spot diameter} = 1.57(\lambda/\text{NA})$$

where λ = wavelength
NA = numerical aperture

Thus, increasing NA decreases spot size and improves system resolution. However, the depth of focus (F_d) of an optical system at visible wavelengths is

$$F_d = \lambda/(2\text{NA}^2)$$

Therefore, the tolerance of the focus will be more critical as NA is increased.

Reference

1. T. Bruegge, from S. Weiss, "Rules of Thumb, Shortcuts Slash Photonics Design and Development Time," *Photonics Spectra*, Oct. 1998, pp. 136–139.

INDEX IMPACTS ON WAVEFRONT ERROR

Wavefront error (WFE) can be calculated as follows:

$$\text{WFE} = \sigma|\Delta n|$$

where   WFE = wavefront error in wavelengths of light at a given wavelength (the HeNe wavelength of 0.63 μm is typically used to characterize optics in the visible wavelengths)
σ = surface error in wavelengths of light
Δn = magnitude of the change in the index of refraction across the interface $(n_1 - n_2)$

Discussion

This equation is a result of geometrical optics, wavefront calculations, and basic physics. This simple rule is valid across the electromagnetic spectrum; it allows one to estimate total wavefront error.

The rule provides guidance in determining a specification for the irregularity (flatness) of an optical surface based on the material's index of refraction and the allowable wavefront error.

This assessment allows cost and producibility trades based on the difficulty in achieving a given surface figure and the material's index. This rule gives the project engineer important ammunition to fire at the optic designer to force a change to a lower index material (or vice versa).

When estimating overall system performance, the quantity that the analyst really desires is the wavefront error, not the surface error. The wavefront error can be related to the surface error by the above rule. One can see that the more an optical material's index varies from that of 1 (the index of refraction for a vacuum and approximately that of air), the more stringent the surface figure needs to be.

Mirrors tend to be the most critical, as their change in index is $n_1 - n_2 = 1 - (-1) \approx +2$. Fortunately for the pocketbook of anyone who wears eyeglasses, ordinary glass is less critical: $n_1 - n_2 \approx 1 - (1.5) = -0.5$. Immersed or buried surfaces hardly matter, as $n_1 - n_2 \approx 1.6 - (1.5) = -0.1$.

Reference

1. Private communications with W. Bloomquist, 1995.

POLARIZER EFFICIENCY

A good polarizer is 95 percent efficient and achieves attenuation of 10^5.

Discussion

Polarizers are frequently used in optical telecommunications to isolate beams, prevent back propagation (resulting in laser "pull"), to limit dispersion, and to calibrate gratings.

Perfection doesn't exist in nature, including polarizers. A polarizer's basic figure of merit, at the specified wavelength, is the loss in the allowed polarization along with the open/shut ratio of two identical polarizers aligned versus crossed. The "open" configuration should have the maximum transmission. The best ones generally available achieve losses of 5 percent or transmission efficiency of 95 percent, in the state in which they are configured to transmit, and open/shut ratios of 100,000.

Hobbs points out that a stack of microscope slides can make a moderately good polarizer. Assuming that the coherence of the light is low enough that etalon fringes can be ignored, n glass plates stacked and oriented at the Brewster angle will attenuate the s-polarized light by 0.8^n. Obviously, stacking 30 or 40 plates can yield excellent "s" polarization attenuation with little loss to the "p," but with multiple reflections and wavefront error.

Reference

1. P. Hobbs, *Building Electro-Optical Systems, Making It All Work,* John Wiley & Sons, New York, pp. 188, 192, 2000.

QUICK ESTIMATE OF DIFFRACTION

The angular diameter of the Airy disk is commonly used to estimate the diffraction-limited spot size and defined as

$$2.44\frac{\lambda}{D}$$

where D = aperture diameter in any linear dimensional units
         λ = wavelength in the same units as D

Discussion

For convenience, in the infrared, this may be adjusted to

$$D_b = 244\frac{\lambda}{D}$$

where D_b = diffraction blur in microradians
         D = aperture diameter in centimeters
         λ = wavelength in micrometers

It is important to note that this accounts for diffraction only; it does not include effects aberrations, misalignments, scatter, and so on.

Both of the above relationships assume a clear circular aperture; it is different for noncircular apertures. They also assume a narrow spectral band so that λ is well defined. If using a broad band, use the longest wavelength.

It should be noted that this quick equation provides a fundamental limit. For unobstructed circular apertures, the above Airy disk description will not be violated; someone claiming better resolution must be using super-resolution and *a priori* target knowledge or synthetic apertures.

The designer of free-space optical communication systems should consider that the central Airy disk is smaller than predicted by the above rule if concentric obscurations are present, as in many reflecting telescopes. In fact, there have been telescopes designed with large central obscurations providing an aperture in the shape of an annulus. This system yields a very small Airy disk. However, the fraction of energy is also less in the Airy disk; more is dumped to the outer rings, which tends to greatly reduce overall MTF. Therefore, there is a reduction in signal strength, which is key for many free-space telecom applications.

For a perfect system, the bright disk (Airy disk) internal to the first dark ring contains about 84 percent of the energy. The remaining is found in low-energy rings around the Airy disk (e.g., the first bright ring surrounding this disk contains an additional 7 percent, the second has 2.8 percent, and the third has 1.5 percent).

WAVELENGTH-TO-FREQUENCY CONVERSION

The frequency bandwidth of a system can be related to the wavelength bandwidth by

$$\Delta f = \frac{c(\Delta\lambda)}{n\lambda^2}$$

where Δf = frequency bandwidth
  c = velocity of light
  λ = center wavelength
  $\Delta\lambda$ = wavelength bandwidth
  n = index of refraction (1 for a vacuum, nearly 1 for air, and typically around 1.5 for the core of a fiber)

Discussion

In telecom, the wavelength bandpass often is expressed as a frequency. Using the above conversion, one can move between frequency and wavelength space.

The wavelength of light and the frequency at which it vibrates is related simply by $f = c/\lambda$.

This can be simplified even more if we assume that λ is 1550 nm and the index of refraction is 1.5; then the above equation becomes

$$\Delta f \approx 83.2 \times 10^{18}(\Delta\lambda)$$

where Δf is expressed in hertz and $\Delta\lambda$ is in meters.

COATINGS AND POLARIZATION

If you are specifying optical coatings and the polarization can be either "s" or "p," it is easiest and least costly to reflect "s" and transmit "p."

Discussion

Given modern optical manufacturing process, optical coatings can more easily be made and tested to reflect the "s" and transmit the "p."

Reference

1. G. Enzor, from Weiss, S., "Rules of Thumb," *Photonics Spectra,* Oct. 1998, pp. 136–139.

TOLERANCING OPTICS

$$T = \sqrt{\sum t^2}$$

where T = total effect
     t = effect of each element's tolerance

Discussion

This equation is an often-used rule for tolerancing all optical designs. It indicates that the probable total effect, T, of a set of statistically independent, normally distributed tolerances is equal to the root sum square of the individual tolerance effects. In other words, the degrading effect on optical performance from the tolerance of a single element can be considered independent from the effects of the tolerance of the other elements. Thus, the total effect is just the RSS of the individual effects.

Thus, if there is one element in an optical train that has a much larger effect than the others, it will dominate, no matter how many other elements there are. Also, if two elements have the same effect, then the combination of the two will degrade performance by only the square root of two.

Reference

1. W. Smith, from S. Weiss, "Rules of Thumb, Shortcuts Slash Photonics Design and Development Time," *Photonics Spectra*, Oct. 1998, pp. 136–139.

ALIGNMENT DIRECTION

To simplify optical alignments, make all adjustments from the same direction.

Discussion

Perform manual and nonmanual adjustments identically (in the same direction) during a measurement run; adjust only for aberrations (image quality) or alignment (image location), and determine reference points in a system.

Reference

1. R. Reiss, from Weiss, S., "Rules of Thumb," *Photonics Spectra*, Oct. 1998, pp. 136–139.

REFRACTIVE INDEX OF LITHIUM NIOBATE

Refractive indices of LiNbO$_3$ can be expressed as simple equations.

Discussion

With λ being the wavelength in microns, the following simple empirical equations approximate the index of refraction (n) for LiNbO$_3$ for wavelengths of telecom interest; e and o refer to the extraordinary and ordinary rays.

$$n_o = 2.195 + \frac{0.037}{\lambda^2}$$

$$n_e = 2.122 + \frac{0.031}{\lambda^2}$$

Reference

1. E. Garmire, "Sources, Modulators, and Detectors for Fiber-Optic Communication Systems," Chap. 4 in *Fiber Optics Handbook,* M. Bass, Ed. in Chief, and E. Van Stryland, Assoc. Ed., McGraw-Hill, New York, p. 4.53, 2001.

CYANOACRYLATE ADHESIVE

Don't use "Super Glue™" near critical optical components or the optical interfaces of network elements.

Discussion

This is a conservative caution about the use of cyanoacrylate adhesive near any other optical component. This type of glue produces a variety of volatile gases that are bad for any type of optics—even eyeglasses. Fogging can result. Testing at elevated temperature in a closed container with a disposable optical component can help identify the risk.

References

1. http://www.spie.org/web/oer/january/jan99/techgroups.html#Super.
2. http://optima-prec.com/notes.htm.

TEMPERATURE EQUILIBRIUM

An optical system is in thermal equilibrium after a time equivalent to five times its thermal time constant.

Discussion

When an optical system is exposed to a different temperature, thermal gradients will cause spacing and tilt misalignments and distort the image. After five thermal time constants, the difference in temperature is less than 1 percent. This is found from Newton's equation for thermal change, which shows that

$$T = T_o e^{-t/t_o}$$

where T_o = initial temperature
t_o = thermal time constant of the system

After five time constants, the temperature has settled to nearly its final value. The difference is less than 0.007.

12

Splicing and Connectors

This brief chapter contains rules relating to the art of spicing and connecting fibers. Providing a termination to a fiber and properly getting the light into the next fiber may seem trivial, but it is of utmost importance. A fiber cable must be spliced every 10 km or so, simply because trucks can't hold spools that hold more. (Underwater cables are spliced every few hundred kilometers, as ships can hold larger reels.) Connects are required at every EDFA and switch, and a packet may traverse numerous connectors at each Telecom hotel. The slightest bit of back-reflection or loss can add up over this link to cause intermittent and permanent failures.

Generally, *connector* refers to a fiber termination that allows the fibers to be connected and disconnected multiple times. On the other hand, a *splice* is a "fusion" or "weld" used to form the connection, and it is meant to be permanent and never disconnected.

The quality of splices and connections can be affected by tolerances on the lateral misalignment of the fibers, angular misalignment, gap lengths, mismatched numerical apertures, residual contamination, and mode mismatch. Also affecting the quality are the tolerances on the fiber geometry itself such as the ovality, curl, diameter, and core/clad concentricity of the fiber. When making a splice or installing a connector, the fatigue of the human and, if outdoors, even wind can play a role in performance, as it can cause vibration in the splicing equipment.

The primary concerns for a splice or connection are the potential for loss of signal and the amount of back-reflection. All of the above-mentioned mechanisms can result in some loss of light. Although this can be trivially small for a single, new connection or splice, it can cascade to disastrous levels for a link with tens or hundreds of connections and splices that are several years old, gathering dust and water, and mechanically flexed.

The loss at a splice is typically 0.1 to 0.5 dB, and getting lower all the time with increased process control, better personnel training, and improved technology. Connectors tend to have higher loss, as they do not physically fuse (melt) the fibers together.

In addition to low loss, another key parameter is reflected light. To put it simply, a connector or splice is a discontinuity in the flow of the light. The trick is to make this discontinuity as small and invisible to the light as possible. However, there is often a change in index due to an air gap or tolerances on the fiber material's index of refraction. Thus, there is some back-reflection, which interferes and shows up as noise. This back-reflection can cause multiple failures. Not only does it decrease the signal-to-noise ratio, but reflections of –30 dB can destabilize laser diodes, and connector splice costs can be a significant portion of a link. To minimize these effects, the designer should use limit coupling, specify good couplers, include optical isolators, and consider off-axis surfaces.

Splices and connectors can also produce a noise phenomenon called *modal noise*. A coherent laser source can experience random coupling between fiber modes. The noise can affect the bit error rate. The noise is more pronounced with multimode fibers and is considered nonexistent for single-mode systems. This is another reason why the telecom industry is almost exclusively deploying single-mode fibers for long links. The rules in this chapter concern the two main issues: loss and reflections. There is a rule that contains several equations for calculating the coupling ratio, and another large rule has several equations describing splice loss. Other rules are presented for loss due to lateral gap and angular misalignment, including fiber mismatches. Another rule deals with cleaning, and empirical field data are presented regarding the transmission gains after cleaning.

FIBER ADAPTER AND CONNECTOR CLEANING

If insertion loss suddenly increases with an adapter or connector that has been used several times, the cause may be dirt.[1]

Discussion

Fiber adapters are used to mechanically and optically connect two fiber optic connectors. Fiber optic adapters are almost always female-to-female, while connectors are male.

Usually, a fiber adapter consists of two main elements: the alignment sleeve and the housing. The alignment sleeve is the critical part of the adapter, providing the alignment of the two-connector ferrules. The alignment sleeve is usually a split "C" and is made from a hard, low-wear material. The housing provides the important mechanical connection that holds everything together. The easiest way of cleaning adapters is with a pipe cleaner or alcohol-wetted cotton.

In 1999, Telcordia Technologies, in cooperation with Ameritech, Bell Atlantic, Bell South, Southwestern Bell, and US WEST, conducted a study of the reliability of field enclosures. They gathered statistics on the benefits of cleaning the fiber connections within those field enclosures. For connections deployed in telecom field applications, highway dirt and humidity degrade the connection over time. Highway dirt is typically 5 to 50 microns in size and tends to work its way into connectors over time. In all cases, they experienced a marked improvement in transmission through the connector after cleaning. [2]

Finally, the problem is not always a dirty adapter, so beware. It may be a broken fiber.

References

1. Product catalog, "Single Mode Fiber Optic Products," 2nd ed., Wave Optics Inc., Mountain View, CA, p. 7, April 1994.
2. R. Zelins, "The Effects of Environmental Exposure on Fiber Optic Connectors in Outside Plant Closures," *Proceedings of the National Fiber Optic Engineers Conference*, 2000.

CONNECTION LOSS IS HIGHER AT SHORTER WAVELENGTHS

The loss in decibels at a connector is approximately 6.5 percent greater at 1300 nm than at 1550 nm.

Discussion

Generally, the loss in a connector is slightly larger at shorter wavelengths. Surprisingly, this is consistent with King's[1] equations, which are presented in the rule, "Coupler Loss as a Function of Gap" (p. 303). These equations have the wavelength as the numerator of the attenuation equations, which obviously indicates that loss increases with wavelength. However, these equations aren't really in "closed form" with respect to the wavelength. Both the mode field radius and the index of refraction also change with wavelength, making attenuation a complex function of wavelength. These effects tend to decrease the loss at longer wavelengths and tend to slightly dominate over the appearance of the wavelength in the numerator.

The above rule was generated by the authors based on measured data given by King and presented in the table below.[2]

Nominal attenuation at 1550 NM (dB)	Measured at 1550 NM (dB)	Measured at 1350 NM (dB)	Percent difference
3	3.1	3.3	6.5
6	5.9	6.3	6.8
9	9.0	9.7	7.8
12	11.7	12.9	10.3
15	14.8	15.8	6.8
18	17.8	18.7	5.1
20	20.1	20.3	1.0

References

1. W. King, J. Malluck, and D. Ridgway, "Plastic-Gap Attenuation," *Proceedings of the National Fiber Optic Engineers Conference*, pp. 742–751, 2001.
2. W. King, J. Malluck, and D. Ridgway, "Modular Adapter-Attenuators," *Proceedings of the National Fiber Optic Engineers Conference*, 2000.

CONNECTION BACK-REFLECTION

A model developed by Duff for the power penalty due to multipath back-reflection gives the following approximation:

$$P = 10 \log(1 - 0.7N\alpha)$$

where P = power penalty due to back-reflection in dB
N = number of reflection points (connections) in the link
α = geometric mean of the connector reflections

Discussion

In addition to transmission loss, one also has to consider the back-reflection that occurs at a change of index, as with a connector. The amount of light that is reflected back from a fiber connection must be minimized to prevent a host of undesirable system and network effects, including increased noise. This is becoming more critical with high data rates and video over fiber links with many connections.

This noise is analogous to multipath interference noise of radio waves in the atmosphere and is often called such in optical telecom. Both noise sources are generated by reflection propagating back and interfering. There is also a secondary effect of increased noise caused by the back-reflection also being reflected and then traveling down the fiber in the original direction but with a time delay as compared with the desired signal.

One way to minimize back-reflections is to ensure that the fibers are in physical contact (touching). One way to ensure this contact is to grind a radius on the ferrule, and another is to angle the ferrule. The angle approach also can ensure that the bulk of the back-reflected energy bounces at an angle such that it doesn't enter the fiber core.

Note that if one imagines N to be very large (e.g., in the hundreds), and if alpha is around 0.01, then the above equation results in taking the log of a negative number, which isn't possible, so the above approximation has some serious bounds.

Given below are some typical back-reflection values for various types of connectors.[2] The measurements below indicate how much lower the back-reflected power is than the transmitted power. Thus, the APC is the best of the above, with the back-reflection being 60 dB less than the transmitted power (although it tends to have the highest loss as well).

	Loss
PC (physical contact)	>30 dB
SPC (super physical contact)	>40 dB
UPC (ultra physical contact)	>50 dB
APC (angled physical contact)	>60 dB

References

1. C. DeCusatis and G. Li, "Fiber-Optic Communication Links (Telecom, Datacom and Analog)," in *Handbook of Optics*, Vol. 4, M. Bass, Ed., McGraw-Hill, New York, pp. 6.14, 2001.

2. M. Peppler, "An Introduction to Fiber Optic Connectors and Splicers," in *Photonics Design and Applications Handbook*, Laurin Publishing, Pittsfield, MA, p. H-162.3, 2000.

3. C. DeCusatis and G. Li, 2001, "Fiber-Optic Communication Links (Telecom, Datacom and Analog)," in *Fiber Optics Handbook*, M. Bass, Ed. in Chief, and E. Van Stryland, Assoc. Ed., McGraw-Hill, New York, p. 6.14, 2001.

COUPLER LOSS AS A FUNCTION OF GAP

The loss (in decibels) as a function of the gap can be approximated by

$$A = 10\log\left\{\left[\frac{2\lambda g}{\pi n d^2}\right]^2 + 1\right\}$$

where A = coupler's attenuation in decibels
  λ = free-space wavelength in nanometers
  g = gap length (or element thickness) in millimeters
  n = index of refraction of the material in the gap
  d = mode field diameter of the fiber in microns

Discussion

In practice, loss from a coupler can range from 0.25 dB for a single-mode coupler to 1.5 dB for large-core multimode couplers.[2] Obviously, the loss in a coupler is strongly dependent on the type of coupler, environmental conditions, coupler design, and tolerances.

The loss is normally defined as the difference between input power and output power. In short, it is a measure of the efficiency of the coupler. The splitting loss is the fraction of light appearing at each output port. The actual loss is determined by adding the excess loss to the splitting loss.

The figure was generated using the above equation and assuming a 10-micron mode-field diameter, 1550-nm light, and an index of refraction of 1.5. Note that the equation naturally handles the convenient (yet unconventional) mix of units including nanometers for the wavelength, millimeters for the gap, and microns for the mode field diameter. Figure 12.1

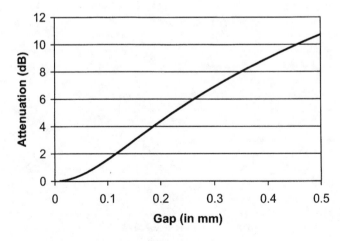

Figure 12.1   Attenuation vs. gap.

illustrates the sensitivity of loss to that of the gap between fibers and underscores the need to keep this gap as small as possible.

The above rule assumes that the fibers are matched, or have very close to the same size core diameter. Otherwise, the coupler has an additional loss due to the mismatch in mode field diameters, and an additional term must be added to the above equation, yielding

$$A = 10\log\left\{\left[\frac{4\lambda g}{\pi(d_1^2 + d_2^2)n}\right]^2 + 1\right\} + 20\log\left[\frac{d_2^2 + d_1^2}{2d_1 d_2}\right]$$

where d_1 and d_2 = two fibers' mode field diameters

References

1. W. King, J. Malluck, and D. Ridgway, "Plastic-Gap Attenuation," *Proceedings of the National Fiber Optic Engineers Conference,* pp. 742–751, 2001.
2. C. Fox, Ed., 1993, Vol. 6, *Infrared and Electro-Optical System Handbook,* ERIM, Ann Arbor, MI, and SPIE, Bellingham, WA, p. 290.

SPLICE LOSS

$$\text{Loss (in dB)} = -10 \times \log\left[\frac{16\,n_1^2\,n_3^2}{(n_1 + n_3)^4} \times \frac{4\sigma}{q} \times \exp\left(-\frac{\rho\mu}{q}\right)\right]$$

where $\rho = (k\omega_1)^2$

$q = G^2 + (\sigma + 1)^2$

$\mu = (\sigma + 1)F^2 + 2\sigma FG \sin\theta + \sigma(G^2 + \sigma + 1)\sin^2\theta$

$F = \dfrac{d}{k\omega_1^2}$

$G = \dfrac{s}{k\omega_1^2}$

$\sigma = \left(\dfrac{\omega_1}{\omega_2}\right)^2$

$k = \dfrac{2\pi n_3}{\lambda}$

n_1 = core refractive index
n_3 = refractive index of medium
λ = wavelength
d = lateral offset
s = longitudinal offset
θ = angular misalignment
ω_1 = $1/e$ mode field of transmitting fiber
ω_2 = $1/e$ mode field of receiving fiber

Discussion

Splice loss results from a combination of several mechanisms, many dealt with as independent issues by other rules in this chapter. The above equation encompasses most of the major loss mechanisms of a splice. The above has been verified using large databases of splices and found to closely represent the real world. Note that n_2 is not present, as this generally refers to the index of refraction of the cladding, which might not be the same as n_3.

Smith points out that a good theoretical model, as the one presented here, can be

...used to draw conclusions about future fiber geometry targets necessary to achieve "blind" splicing capability where no testing would be required due to the level of performance of the resulting splice distribution. We will define "blind" as the ability to achieve splices with a risk of exceeding a targeted specification of less than 1000 ppm with no testing and remating. With an excellent core/clad concentricity distribution characterized by a C_{pk} of 1.3 against the current industry specification of 0.8 μm, and with excellent curl distributions, users should expect blind splicing capability against an individual splice loss of 0.3 dB and an average splice loss of 0.12 dB. ...the lowest practical blind splicing capability would be about 0.2 dB for individual splices and 0.08 dB for average splices.[1]

This rule can also apply in some circumstances to connectors and coupling between waveguide devices if $\omega_1 = \omega_2$.

Reference

1. G. Smith, "A Numerical Analysis of Splice Loss Dependence on Optical Fiber Geometry," *Proceedings of the National Fiber Optic Engineers Conference,* 1997.

MODE FIELD DIAMETER MISMATCH LOSS

The loss associated with mode field diameter (MFD) mismatch between single-mode fibers is

$$
\text{Loss [dB]} \approx -10\log\left[\frac{4}{\left(\dfrac{MFD_1}{MFD_2} + \dfrac{MFD_2}{MFD_1}\right)^2}\right]
$$

where MFD_1 = mode field diameter of one of the fibers
       MFD_2 = mode field diameter of the other fiber

Discussion

Differences in the mode field diameter between single-mode fibers (of the same type or different types) lead to a signal loss. The above equation can be used to estimate the loss and can be used with the measured or specified MFD, or the uncertainty of the MFD. The user should realize that MFD is a function of wavelength and changes across the C and L bands, as illustrated in Fig. 12.2.

Farley et al.[1] point out, "There are several companies manufacturing dispersion devices. Each company should supply MFD and A_{eff} specifications as a function of wavelength to their customers. This will aid the deployment of these components. Using the goniometric radiometer technique,

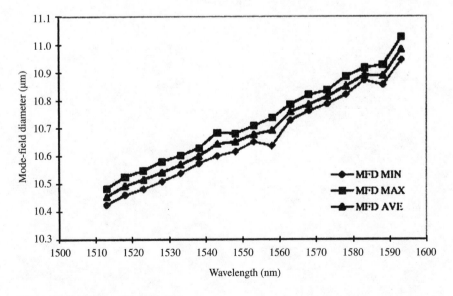

Figure 12.2   MFD loss (from Ref. 1).

both fast and accurate measurements can be made. Where other methods may take hours, this method takes less than 20 s."

Farley et al. also point out that, as an example, the uncertainty of 10 percent of MFD in a typical single-mode fiber results in a loss uncertainty of about 0.039 dB or 0.9 percent, while an uncertainty of 0.5 percent leads to a loss uncertainty of merely 0.00011 dB or 0.0025 percent.

The MFD is also related to the effective area of the fiber core (A_{eff}), which in turn is related to the nonlinear index n_2, and (n_2/A_{eff}) defines the nonlinear effects in the fiber. Nonlinear effects tend to be less pronounced for larger-diameter fibers.

References

1. H. Farley et al., "Mode-Field Diameter and Effective Area Measurement of Dispersion Compensated Optical Devices," *Proceedings of the National Fiber Optic Engineers Conference*, 2000.
2. Optics for Research, fiber optics product catalog, Caldwell, NJ, 2001, and http://www.ofr.com/ 2002.
3. Staff of Lucent Technologies, "Mixing TrueWave RS Fiber with Other Single-Mode Fibers in a Network," http://www.ofsoptics.com/resources/mixing_fibers4.pdf, 2002.

LOSS DUE TO ANGULAR MISALIGNMENT OF A SPLICE

For single-mode fibers,

$$\text{Loss (dB)} = -10\log e^{-\left(\frac{n\pi\omega(\sin\theta)}{\lambda}\right)^2}$$

where n = fiber's index of refraction
   ω = mode field radius
   θ = angle of misalignment (0° is perfect alignment, and 90° is orthogonal)
   λ = wavelength of light

And, for multimode fibers,

$$\text{Loss (dB)} = -10\log\left[1 - \frac{8n(\sin\theta)}{3\pi(NA)}\right]$$

where NA = numerical aperture

Discussion

Splice loss is a combination of several mechanisms, with angular alignment being one of the largest. If the fibers are not perfectly parallel, there will be increased losses due to difficulty in relaunching the light into the fiber. Even a slight angle between the fiber ends will result in a mismatch in numerical aperture.

The above equations assume that the fibers are of the same type and have identical mode field radii, diameters, indices of refraction, and numerical apertures, and uniform MFD.

Splices, being more permanent, have less loss than do connectors.

References

1. C. Good et al., "Improvements in Splicing Polarization-Maintaining Fiber," *Proceedings of the National Fiber Optic Engineers Conference*, pp. 362–369, 2001.
2. Private communications with D. McCal and G. Fanning, 2002.
3. http://www.ofr.com/tools, 2002.

EXTINCTION RATIO DUE TO ROTATIONAL ANGULAR MISALIGNMENT OF A SPLICE

$$ER \text{ (dB)} = 10 \log (\tan^2\alpha)$$

where $ER =$ extinction ratio (theoretically infinite for 90°)
$\alpha =$ incident angle between (as shown in Fig. 12.3) the fiber's
transmission axis and the angle of the light input

Discussion

The reader is cautioned that there are several "extinction ratios" used in the vernacular. This extinction ratio is the ratio of the power traveling in the slow versus the fast axis, or the maximum light intensity versus the minimum light intensity measured 90° from the maximum. This is commonly called the ordinary and extraordinary polarizations when discussing the birefringence of optical materials. The coupling from one fiber to another depends on the polarization state of each. In terms of the ratio, an angular misalignment causes less light of the measured polarization to be successfully launched and may also allow leakages of the other polarization. This results in a reduction of the difference in intensity between the maximum and minimum. The rapid decrease in the extinction ratio is shown graphically in Fig. 12.4 using both the above equation and measured data. The angular misalignment should be kept small to minimize this degradation.

Practically speaking, a polarization mode (PM) fiber can be aligned to another PM fiber (of the same sort) in an automatic splicing machine to

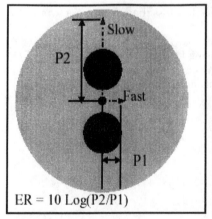 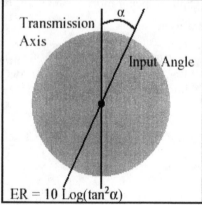

Figure 12.3   Extinction ratio due to angular offset.

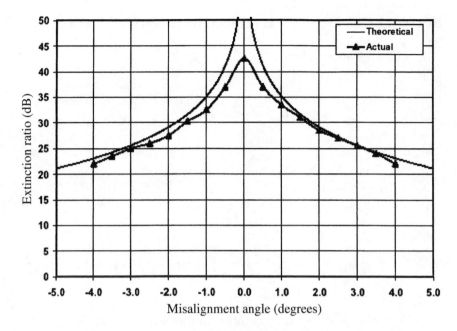

Figure 12.4 Extinction ratio for various degrees of misalignment (from Ref. 1).

–35 dB at best. Manufacturers claim that they typically have a variation of <3 dB, but it is hard to verify these claims.

Additionally, a ±5° alignment error is easy to achieve during production by aligning a fiber with the appropriate axis of a modulator. This is done by visually aligning the stress rods to a target plane. It is not possible to do better than approximately ±2° visually; this requires making an extinction measurement while rotating the fiber. Although possible in the laboratory, this is a tricky measurement to conduct on each piece of hardware in production.

References

1. C. Good et al., "Improvements in Splicing Polarization-Maintaining Fiber," *Proceedings of the National Fiber Optic Engineers Conference*, pp. 362–369, 2001.
2. Private communications with D. McCal and G. Fanning, 2002.

SPLICE LOSS DUE TO LATERAL ALIGNMENT ERROR

1. For a single-mode fiber,

$$\alpha_{lat} = -10\log e^{-\left(\frac{d}{\omega}\right)^2}$$

where α_{lat} = loss due to lateral offset in decibels
ω = mode field radius (assuming both fibers have the same)
d = lateral offset

or

$$\alpha_{lat} = -10\log\left[\left(\frac{2\omega_1\omega_2}{\omega_1^2 + \omega_2^2}\right)^2 \times \exp\left(\frac{-2d^2}{\omega_1^2 + \omega_2^2}\right)\right]$$

where α_{lat} = loss due to lateral offset in decibels
ω_1 = mode field radius of fiber one
ω_2 = mode field radius of the other fiber
d = lateral offset in microns

2. For multimode fibers,

$$\alpha_{lat} = -10\log\left[1 - \frac{8d}{3\pi a}\right]$$

where a = fiber radius

Discussion

Splice loss results from several mechanisms (see other rules in this chapter). Among the mechanisms for loss are lateral misalignment (as shown in Fig. 12.5), angular misalignment, gap lengths, mismatched numerical apertures, residual contamination, and mode mismatching. Generally, the dominant sources are lateral and angular misalignments.

As nothing can be aligned perfectly, alignment tolerances must be allowed for in manufacturing and network design. Generally, one fiber to another can be aligned to within 1/3 to 2/3 of a micron as illustrated by Fig. 12.6.

The first two equations above are valid for single-mode fibers, and the final equation is valid for multimode fibers of the same radius.

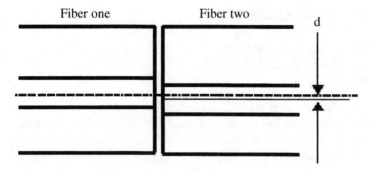

Figure 12.5   Lateral offset.

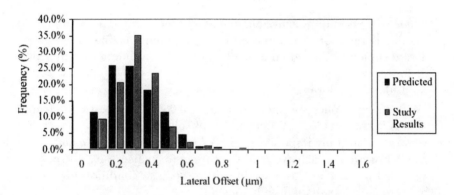

Figure 12.6   Representative statistics of alignment offset from the large loss laboratory study.

References

1. S. Cooper and R. Erskine Jr., "Practical Guidelines for Mass Splicing," *Proceedings of the National Fiber Optic Engineers Conference*, 2001.
2. http://WWW. OFR.com/tools, 2002.

SPLICE LOSS IN DECIBELS

A typical splice loss is about 0.08 to 0.2 dB and may approach 0.04 dB with special care and active monitoring approaches.

Discussion

Several rules in this chapter address a mathematical model to calculate splice loss. The above rule and supporting data in Figs. 12.7 and 12.8 indicate the achievable splice loss. A multitude of error sources can add up to make a very low-loss splice, or their mix can result in a very lossy splice. This rule applies to the most popular type, which is the fusion splice. Generally, there is a distribution of these error sources that results in a distribution of losses.

Some splicers employ active alignment and measurement techniques to allow the user to more accurately position and orient the fibers for a splice. These techniques tend to be difficult and time consuming, but they provide for lower losses.

Consider the following:

- "With a single splice specification limit of 0.08 dB, the best yield achievable is 91 percent (i.e., 9 percent of all splices will require revisit and rework). If we only impose a 0.08-dB specification on the *average* span splice loss, we can achieve 100 percent conformance (i.e., no rework), provided we have at least six joints in the span (a condition that would apply to most long-haul installations)."[1]
- Studies indicate that a 0.08-dB loss splice can be achieved about 80 percent of the time, but a 0.03-dB loss less than 20 percent of the time for a LEAF-96.[1]
- Field data that indicate that it takes between two and three times longer to rework a 0.1-dB splice than it does a 0.2-dB loss splice. So the user should beware and include the labor for the initial installation and rework before specifying a low-loss splice.[1]
- Theoretical analysis indicated that the best practical splice using conventional techniques will be approximately a 0.08-dB loss specification for a large percentage of the splices.[2]
- In a 12f ribbon LEAF to LEAF, the average splice loss was 0.07 dB with a standard deviation of 0.06 dB.[3]
- With V-groove splicers, the average loss from nonzero-dispersion (NZD) fibers is twice that of standard single-mode fibers.[4]
- Mean splice loss improved from approximately 0.172 to 0.044 dB from 1985 to 1995.[4] This improvement is graphed in Fig. 12.7.
- T-connectors have higher loss; typical connector loss values average about 0.5 dB worst case for multimode and slightly higher for single mode.[5]

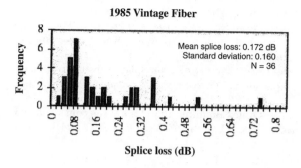

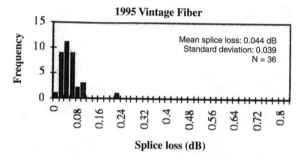

Figure 12.7   Improvements in fiber quality and splicing techniques have led to smaller splice losses (from Ref. 4).

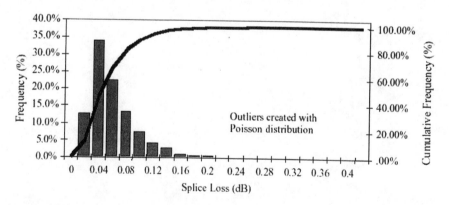

Figure 12.8   Representative splice loss (from Ref. 3).

References

1. D. Gibb et al., "Setting Reasonable and Practical Splice Specifications for Field Installation of Single-Mode Fiber Cables," *Proceedings of the National Fiber Optic Engineers Conference*, pp. 379–390, 2001.

2. G. Smith, "A Numerical Analysis of Splice Loss Dependence on Optical Fiber Geometry," *Proceedings of the National Fiber Optic Engineers Conference*, 1996.
3. S. Cooper and R. Erskine, "Practical Guidelines for Mass Splicing," *Proceedings of the National Fiber Optic Engineers Conference*, 2001.
4. D. Duke et al., "Migration of Core-Alignment Fusion Splicing from the Splicing Van to Outdoor/FTTH Applications," *Proceedings of the National Fiber Optic Engineers Conference*, 2000.
5. C. DeCusatis and G. Li, "Fiber Optic Communication Links (Telecom, Datacom and Analog)," in *Handbook of Optics*, Vol. 4, M. Bass, Ed., McGraw-Hill, New York, pp. 6–8, 2001.

13

Laser Transmitters

Revolutions in optics are infrequent but, in the case of lasers and laser diodes, the impact was immediate and profound. Although it took the conjunction of the invention of many complementary technologies [lasers, fibers, detectors, erbium-doped fiber amplifiers (EDFAs), and so forth] to make large optical telecom networks possible, much of the success depends on the ability to cheaply and efficiently create the transmitters that generate and repeat the modulated beams of light that carry information.

In addition to enabling communication systems, lasers have found their way into every home, most cars, and even people's pockets. For the more experienced (older) designer, it is astonishing to think that, just a few decades ago, lasers were expensive, huge, and hard to align. More important, the wide use of lasers has meant that experience with them can be transferred to the fiber community, the free-space communications community, and so on. In fact, much of the theory for both applications has been around for years and is the emphasis of this chapter.

Not surprisingly, the rules covered in this chapter concentrate on the ability of particular devices to create light for communication systems and methods for coupling that light into fibers. It also includes information on the noise properties of various types of lasers, including the currently popular EDFA systems. EDFA gets attention throughout the chapter because of its profound impact on fiber system design. To be complete, we have included a couple of rules related to the temperature sensitivity of EDFAs and management of those devices to the desired performance for a range of operating conditions. Fiber Bragg grating systems are also addressed because of their growing importance in supporting the needs of dense wavelength division multiplexing (DWDM) due to their ability to define narrow wavelengths.

For free-space communication applications, we have included rules related to beam propagation and properties of such beams as a function of the transmitting aperture.

The largest rule in this chapter is the one dealing with eye safety. This is justified by the large number of people who will be exposed to laser radiation during installation and repair of fiber systems. The reader is encouraged to check the appropriate Telcordia, ANSI, OSHA, and FDA specifications.

The reader has access to a wide variety of texts that describe both the physics and applications of lasers. Siegman[1] provides the current standard of excellence for a sophisticated presentation of laser physics and design approaches. It is a great resource and should be accessible to everyone who uses or expects to use lasers. Less complex texts are also available in most college bookstores. For the entry-level student, the laser industry can be a great resource. For example, optical manufacturer Melles Griot includes a great deal of useful information in its product catalog. In addition, new laser users will want to look at various magazines, such as *Laser Focus World,* because they complement the much more complex presentations found in journals such as *Applied Optics* and *IEEE Journal of Quantum Electronics.* Optics books should not be overlooked, because most provide a pretty thorough description of laser operation and applications and provide additional references for consideration.

Reference

1. A. Siegman, *Lasers,* University Science Books, Mill Valley, CA, 1986.

ACOUSTO-OPTIC MODULATION FREQUENCY AND RISE TIME

The modulation frequency of an acousto-optic device can be stated as

$$F_m \tau = 0.75$$

and the optical rise time as

$$T_r = 0.64\tau$$

where F_m = optical intensity modulation frequency where the depth of modulation is -3 dB
  $\tau = D/V$
  D = optical beam diameter
  V = sound velocity inside the acousto-optic material

Discussion

When the depth of modulation is -3 dB, or 50 percent, as defined above, the intensity contrast is 3:1. Variable τ is known as the access time or the transit time of an acoustic wavefront across an optical beam of diameter D inside the acousto-optic medium.

These rules assume the classical definition of optical rise time, which is the time it takes for the intensity of an optical pulse to rise from the 10 to 90 percent point when a radio frequency step function is applied to the acousto-optic device. This also assumes that the device's electrical bandwidth is wide enough to pass the Fourier pulse components. Finally, this rule applies to uniform circular beams only.

Reference

1. J. Lekavich, from S. Weiss, "Rules of Thumb, Shortcuts Slash Photonics Design and development Time," *Photonics Spectra,* Oct. 1988, pp. 136–139.

APERTURE SIZE FOR LASER BEAMS

When a Gaussian beam encounters a circular aperture, the fraction of the power passing through is equal to

$$1 - e^{\frac{-2\,a^2}{\omega^2}}$$

where a = radius of the aperture
ω = radial distance from the beam's center to the point at which the beam intensity is 0.135 of the intensity at the center of the beam

Discussion

The beam intensity as a function of radius is

$$I(r) = I_o \exp \frac{-2\,r^2}{w^2}$$

where w = as defined as above
r = radius at which the intensity is specified

The power as a function of size of the aperture is computed from

$$\frac{P(a)}{P_o} = \int_0^a I(r)\frac{2\pi}{P_o}dr = 1 - \exp\left(\frac{-2\,a^2}{w^2}\right)$$

At the $1/e^2$ point of the beam, the intensity is down to 0.135 of the intensity at the center of the beam. Thus, we see that an aperture of $3w$ transmits 99 percent of the beam.

This rule applies to beams that are characterized as Gaussian in radial intensity pattern. While this is nearly true of aberration-free beams produced by lasers, there are some minor approximations that must be accommodated for real beams.

This rule is particularly nice for estimating the size of the aperture needed to pass an appropriate part of a beam.

Of course, as in any system in which an electromagnetic wave encounters an aperture, diffraction will occur. The result is that, in the far field of the aperture, one can expect to see fringes, rings and other artifacts of diffraction superimposed on the geometric optics as a result of a Gaussian beam with the edges clipped off.

References

1. H. Weichel, *Laser System Design*, SPIE course notes, SPIE, Bellingham, WA, p. 38, 1988.
2. A. Siegman, *Lasers*, University Science Books, Mill Valley, CA, p. 666, 1986.

BEAM QUALITY

Beam quality is defined as

$$BQ \approx \exp\left(\frac{1}{2}(2\pi\,WFE)^2\right) = \frac{1}{\sqrt{S}}$$

where BQ = wavefront error (unitless) WFE expressed in waves
S = Strehl ratio[*] (unitless)
WFE = wavefront error

Discussion

This is an approximation that works well when the wavefront error is much less than a wavelength. The beam focusing and collimation that can be obtained is derived from the diffraction theory for plane and Gaussian beams encountering sharp-edged apertures, as described in virtually every optics and laser book. In those analyses, it is assumed that the wavefront is ideal and that there are no tilt or higher-order aberrations in the phase front. This parameter has been developed to deal simply with the additional impact of nonuniform phase fronts in those beams. In a wide variety of applications, this definition is used to define the additional spreading that will be encountered in focused or parallel beams.

There are many other definitions of beam quality, so the reader is cautioned to understand what is meant by BQ in a particular application.

These characterizations of a laser beam are effective measures when the beam is nearly diffraction limited ($S > 85$ percent). For highly aberrated beams, it may be difficult to establish the beam quality. For example, using the definition that relates beam quality to beam size, a highly aberrated beam will have an ill-defined diameter that varies with azimuthal angle, thus limiting the usefulness of the definition.

This rule provides a simple parameter that defines the beam spread of a laser beam. For example, the dimension of the spot of a beam will be expressed as $(\lambda/D)\ BQ$, rather than the ideal λ/D. This means that the spot will cover an area that is proportional to BQ^2. Therefore, we find that the energy density in the beam will depend on $1/(BQ^2)$.

This parameter defines a shortcut to characterize how much a beam deviates from the ideal, diffraction-limited case. As is mentioned above, this can be an effective definition if the BQ is close to unity. For very poor beam quality, such as might result from turbulence and other atmospheric ef-

---

[*]Strehl ratio is the ratio of the on-axis intensity of a beam to the intensity produced by a perfect optic. A similar form of the definition uses the amount of energy falling within the first Airy ring of an optic compared with the energy in the first ring of an aberration-free system.

fects, the entire concept of a well defined beam becomes useless, and this definition fails to characterize the beam.

BQ can also be defined in terms of the power inside a circle at the target.

$$BQ = \sqrt{\frac{P_{ideal}}{P_{actual}}}$$

where the powers are compared within a common radius from the center of the target.

The effect of beam quality is included in the typical diffraction spreading of a beam by

$$\theta_D = \left(\frac{2BQ\lambda}{\pi D}\right)$$

where BQ is the beam quality at the aperture. When BQ is unity, the diffraction angle is

$$\theta_D = \left(\frac{2\lambda}{\pi D}\right)$$

Thus, we see that BQ is included as a linear term for estimating the beam spread of a laser.

Reference

1. G. Golnik, "Directed Energy Systems," in Vol. 3, *Emerging Systems and Technologies*, S. Robinson, Ed., *The Infrared and Electro Optical Systems Handbook*, J. Accetta and D. Shumaker, Ex. Eds., ERIM, Ann Arbor, MI, and SPIE, Bellingham, WA, p. 472, 1993.

Cavity Length and Modulation of Laser Operating Frequency

A change in cavity length produces a change in resonant frequency approximated by

$$\Delta v \approx -v_n(\Delta L/L)$$

where Δv = change in lasing frequency
   v_n = original lasing frequency
   ΔL = change in laser cavity length
   L = original laser cavity length

Discussion

Laser physics shows that the round trip time for a cavity of dimension L must satisfy phase conditions so that amplification can occur. Waves propagating in the cavity that do not match the phase requirements will not be amplified and will be suppressed by ever-present cavity losses. Increasing the cavity length actually reduces the frequency at which the phase condition is satisfied, which is equivalent to saying that long wavelength waves will then be preferentially amplified.

The oscillation (or lasing) frequency of a laser is adjustable over a narrow range by changes in the length of the cavity. Often, one of the cavity end mirrors is mounted on a piezoelectric or magnetostrictive material that allows a small change in length. This change in cavity can be related to the resultant change in frequency by the above equation.

Since frequency and wavelength are directly proportional, the change in wavelength may be found by substituting the original wavelength for the frequency, except that, in the case of wavelength, the minus sign disappears.

To be effective, the change in length must be small compared to the total cavity length. The modulation range is limited to the spacing between modes for a multimode laser. The change in cavity length also impacts the separation in frequency between longitudinal modes and can be used to select only one frequency mode (single longitudinal mode) from the lasing gain envelope inside the laser cavity. The separation between two longitudinal modes is defined as $c/2L$.

References

1. W. Pratt, *Laser Communication Systems*, John Wiley & Sons, New York, p. 67, 1969.
2. A. Siegman, *Lasers*, University Science Books, Mill Valley, CA, p. 41, 1986.

A POWER INCREASE CAN HELP REDUCE THE EFFECTS OF CHROMATIC DISPERSION

The impact of chromatic dispersion in single-mode fibers can be overcome with a power increase

Discussion

Modal dispersion is not a problem in single-mode fibers but chromatic and polarization dispersion can limit performance for long distance transmission. The following equation shows how much additional power must be invested to avoid intersymbol interference caused by chromatic dispersion only. By proper design the penalty can be kept below 1 dB.

$$P_D = 5\log[1 + 2\pi(BD\Delta\lambda)^2 L^2]$$

where P_D is in decibels.

Consider an example where B (bit rate) is 1.25 Gb/s, the fiber length (L) is 1000 km, the chromatic dispersion (D) is 1 ps/nm/km, and the bandwidth ($\Delta\lambda$) is 0.007 nm. For this case, with an admittedly small dispersion, the power penalty is 0.085 dB. This result is achieved by closely matching the operating wavelength to the zero dispersion wavelength to achieve a chromatic dispersion of 1 ps/nm/km.

Reference

1. C. DeCusatis and G. Li, "Fiber-Optic Communication Links (Telecom, Datacom, and Analog)" Chap. 6 in *Fiber Optics Handbook*, M. Bass, Ed. in Chief, and E. Van Stryland, Assoc. Ed., McGraw-Hill, New York, p. 6.11, 2001.

COUPLING OF LEDS

Coupling efficiency (η) from an LED to a fiber is approximated by $(NA)^2$ for $r < a$ and $(a/r)^2(NA)^2$ for $r > a$. Variable r is the radius of the LED, and a is the radius of the fiber.

Discussion

Many old legacy transmitters do not employ lasers, but use light emitting diodes. LEDs are not coherent sources and radiate into a large solid angle. Coupling the light from an LED into a fiber can be an inefficient process. The best solution is to make sure that the fiber is larger than the LED. Large-core, multimode fibers enjoy some success in the industry because of the ease with which LEDs can be matched to them. Solutions of this type can function at data rates exceeding 100 Mb/s.

LEDs also suffer from lack of spectral purity, which limits their application in systems that operate with more than one color per fiber. Finally, the use of large-core fibers in conjunction with LEDs always leads to dramatically higher attenuation and dispersion than is found in designs that use laser sources and single-mode fibers.

Another method to couple light inside a fiber is the use of special lenses such as GRIN (GRadient INdex) lenses designed for that purpose.

Reference

1. I. Jacobs, "Optical Fiber Communication Technology and System Overview," Chap. 2 in *Fiber Optics Handbook*, M. Bass, Ed. in Chief, and E. Van Stryland, Assoc. Ed., McGraw-Hill, New York, p. 2.5, 2001.

EXTINCTION PENALTY

Laser extinction ratio is the ratio between the unmodulated optical power and the modulated power. The reader should take care, as this is sometimes presented in exactly the inverse form. For example, some references define the extinction ratio as

$$\frac{\text{Power}(1)}{\text{Power}(0)}$$

which means that the ratio is the power when sending a 1 compared to the power when sending a 0.

Discussion

Typically, a modulated laser's extinction penalty approaches that of unamplified systems such that

$$\text{Extinction penalty} = 10 \log\left[\frac{1 + r}{1 - r}\right]$$

where r = extinction ratio of the laser given by

$$r = \frac{P_{thr} + \eta_e(I_{bias} - I_{thr})}{\eta_e(I_{mod} + I_{bias} - I_{thr})}$$

where P_{thr} = laser spontaneous emission when the laser is biased at threshold
η_e = slope efficiency
I_{bias} = OFF state laser current
I_{thr} = lasing threshold
I_{mod} = ON state laser current

Semiconductor lasers emit low-power-level light when powered by voltages below their lasing thresholds, and this is called *spontaneous emission.* When used in data transmission, this spontaneous emission results in a background level of unmodulated light. The extinction ratio can also have a component for electrical reflections from driver-to-laser impedance mismatch, which may cause some light emission when in the OFF state. These result in the generation of unmodulated light of the desired wavelength.

The extinction ratio is basically the ratio between the desired modulated light and the unmodulated spontaneous emission, and the "penalty" is log of the resulting "contrast." Extinction ratios for directly modulated lasers are from 9 to 14 dB, whereas externally modulated lasers (with Mach–Zehnder modulators) can exceed 15 dB.

References

1. D. Fishman and B. Jackson, "Transmitter and Receiver Design for Amplified Lightwave Systems," in *Optical Fiber Telecommunications*, I. Kaminow and T. Koch, Eds., Academic Press, San Diego, CA, p. 75–76, 1997.
2. D. Duff, "Computer-Aided Design of Digital Lightwave Systems," *J. Select. Areas Communications*, 2, pp. 171–185.
3. C. DeCusatis and G. Li, "Fiber-Optic Communication Links (Telecom, Datacom, and Analog)," Chap. 6 in *Fiber Optics Handbook*, M. Bass, Ed. in Chief, and E. Van Stryland, Assoc. Ed., McGraw-Hill, New York, p. 6.14, 2001.

TEN RULES OF EYE SAFETY

Everyone needs to take eye safety seriously. To help in this regard, we offer the following obvious (but often overlooked) methods of protection. These are in no particular order of importance, since each can be a factor in the overall exposure. But here are the key methods of maintaining eye safety.

1. Limit laser output power to only what is essential for performing the communication.

2. Use multiple transmission sources.

3. Minimize access to the laser.

4. Display proper eye safety labels.

5. Provide visible indication of laser on/off status.

6. Provide for remote power interlock and take precautions that the interlock cannot be overridden.

7. Properly locate system controls.

8. Use safe alignment procedures.

9. Train users on proper setup and maintenance procedures.

10. Finally, remember the ultimate laser safety rule used world over by laser jocks: "Don't look into the laser with your *good* eye."

Discussion

In normal operation, systems are safe, since all of the light is confined inside the fiber and/or components. However, technicians and installers must be aware of the various standards and methods for eye protection, because the likelihood of laser light exposure can be high during installation, repair, test, and research, or when a fiber has been damaged. Technicians also need to be aware of the relative hazards of various wavelengths to which they might be exposed. For example, 1.3-micron systems can be dangerous, but 1.55-micron systems are safer at the same power levels.

Therefore, technicians and installers exposed to radiation emitted from fibers and other components need to understand safety issues, take proper precautions, and be ready to explain these issues to co-workers. The central issue in working with laser devices is not their inherent power per communications channel, which can be quite low, but the summation of many DWDM wavelengths and the transmission and susceptibility of components of the eye.

In the visible region ($\lambda \approx 400$ to about 700 nm) and part of the near-infrared region ($\lambda = 700$ to 1400 nm), which includes the popular C and L bands, the retina is at risk, because the components of the eye, including the aqueous humor, are transparent. In rare cases, an individual will be sensitive outside these bands. Most commonly, this includes people with heightened UV or near IR sensitivity. These people may be more at risk, since they have increased response from their retinas. Even at longer wavelengths ($\lambda > 1400$ nm), the surface of the eye can be burned.

U.S. manufacturers work under rules and regulations created by the Center for Devices and Radiological Health (CDRH), which is a component of the Food and Drug Administration (FDA).

The CDRH regulation requires manufacturers to place each laser product into one of four classes, depending on its hazard potential. These classes are described below. Note that U.S. and European standards vary somewhat in each class, so be sure to note who issued the certification of the laser to ensure that you are using it safely.

Class 1 includes lasers with sufficiently low power that, under normal conditions, they cannot emit a hazardous level of optical radiation. Because of their inherent safety, no warning label or control measures are required by the CDRH. It is always prudent, however, to avoid exposure to any laser light, if possible.

Class 2 has two subclasses: Class 2 and Class 2a. Class 2 lasers emit in the visible spectrum ($\lambda = 400$–700 nm). These products differ from Class 1 devices in that safe exposure levels are maintained as long as exposures last less than 1/4 s. This value is derived from the reaction time of the blink response that occurs in most people when exposed to a bright light. In general, the blink protects the eye. However, one cannot rely on the blink effect in optical telecom, as the wavelengths are typically longer than the visible response, so the "blink response" is never triggered. Prolonged exposure to a Class 2 laser is a threat to eye safety. Class 2 lasers have power levels less than 1 mW. Class 2a lasers are special-purpose lasers whose power output is less than 1 mW, and they are therefore safer than Class 2 lasers. They are sometimes used in bar-code readers.

Class 3 also has two subclasses. Class 3a products produce only visible light, and these devices cannot injure the unaided eye of a person who reacts with the normal aversion response to a bright light. They are a threat to those who look directly into the beam with a magnifying eyepiece. Class 3a lasers have power levels of 1–5 mW. Class 3b lasers emit over the full spectrum (UV to far infrared) and can cause accidental injury if viewed directly by the unaided eye. They are also hazardous if used as the source of specular reflections. They produce output in the range of 5–500 mW continuous wave (cw), or less than 10 J/cm^2 for a 1/4-s pulsed system.

Class 4 lasers include all lasers with power levels greater than 500 mW cw, or greater than 10 J/cm^2 for a 1/4-s pulsed system. Because of their high power levels, they can threaten health in many different ways. Both burns and eye injuries are possible from these lasers. Obviously, they are a threat, whether exposure is through direct or reflected light.

The maximum permissible exposure (MPE) indicates whether the amount of light is safe. For laser light between 1400 and 1500 nm and exposure times of 0.001 to 10 s, the MPE is $5600t^{0.75}$ W/m^2 (t = time in seconds). For light from 1500 to 1800 nm, and times from 1 ns to 10 s, the MPE is 10,000/t W/m^2. Many standards are widely used to assure safety in the use of these products. They include the following:

1. Laser Product Performance Standard Code of Federal Regulations, Title 21, Subchapter J, Part 1040, Federal Register, 50(161), Tuesday, August 20, 1985, p. 33697.

2. American National Standard for the Safe Use of Lasers, American National Standards Institute Z136.1 (New York: ANSI, 2000).

3. IEC TC76 (Optical Radiation Safety and Laser Equipment) addresses safety aspects of lasers in IEC 60825-2 (Safety of Optical Fiber Communication Systems, 2nd ed.) specifically in relation to tissue damage.

4. IEC TC31 discusses the risk of ignition of hazardous environments by radiation from optical equipment, 2000.

5. ITU-T Question 16/15 is revising Recommendation G.664; Optical Safety Procedures and Requirements for Optical Transport Networks. Thermal damage of fibers, components, and devices caused by high optical intensity are also discussed.

6. IEC 60825-1, Safety of Laser Products—Part 1: Equipment, classification, requirements and user's guide, ed. 1.1, 1998.

7. ANSI Z136.2, American National Standard for Safe Use of Optical FIber Communication Systems Utilizing Laser Diode and LED Sources, 1997.

8. The IEC TC86 SC86C WG3, on optical amplifiers, will also develop a guideline of safety for high optical intensities from fiber-optic amplification.

A selection of key definitions are to be included in the guideline such as maximum permissible emission (MPE), accessible emission limits (AEL), automatic power reduction (APR), restricted area, controlled area and, additional explanations where necessary.

An example illustrates how to compute the tolerable exposure for a Class 1 laser. This derives from FDA Center for Devices and Radiological Health 21 CFR Ch. 1 (4-1-98 ed.), Sec. 1040.10 guidelines, which specify that, for Class 1, "No more than 0.195 mW of average power may pass through an aperture 7 mm in diameter at a distance of 20 cm from the end of the fiber."

Lemoff and Buckman[2] provide an example calculation of eye exposure. We find that a typical mode field diameter is around 9.8 μ. If one assumes a Gaussian beam, the $1/e^2$ radius is 4.9 μ at the end of the fiber. At 20 cm from the end of the fiber, the beam radius is

$$w(z) = \frac{\lambda z}{\pi w_0} = 17 \text{ mm}$$

where w_0 = aperture radius

$z = 20$ cm

The radial distribution of power is

$$I(r) = \frac{2P}{\pi w^2} e^{-2r^2/w^2} \propto e^{-0.0069 r^2}$$

This profile allows us to compute how much of the beam passes through the 7-mm aperture (approximating the eye's pupil diameter). To do so, we must integrate the cylindrically symmetric Gaussian beam.

$$\eta = \frac{2}{\pi w^2} \int_0^a 2\pi r e^{-2r^2/w^2} dr = 1 - e^{-2a^2/w^2}$$

In this case, a = the aperture radius (in this case, 3.5 mm), and w is 17 mm. This leads to a result of $\eta = 0.081$. To compute the power in dBm, we compare the transmission through the 7-mm aperture with the standard of 0.195 mW.

$$P = \frac{0.195 \, mW}{0.081} = 2.41 \, mW = 3.8 \, dBm$$

Thus, a power of level of 2.41 mW or 3.8 dBm is allowed in the fiber.

We also point out that light exposure is not the only threat from fibers. They are, after all, pieces of glass, so one should take care to dispose of scraps carefully to avoid them getting into eyes. In addition, any system that exploits free-space laser communications is an obvious threat, even at so-called "eye safe" wavelengths. With sufficient power, the surface of the eye and skin can be damaged.

References

1. G. Clark, H. Willebrand, and B. Willson, *Free Space Optical Laser Safety,* Light-Pointe Communications, Inc., available at http://www.lightpointe.com/index.cfm/fuseaction/technology.WhitePapers.
2. B. Lemoff and L. Buckman, "WWDM Transceiver Update and 1310 nm Eye Safety," IEEE 802.3 HSSG Meeting, Montreal, Quebec, July 5–9, 1999.
3. Telecommunications Industry Association IEC TC76: Optical Radiation Safety & Laser Equipment. Details at http://www.tiaonline.org/standards/tag/committee.cfm?tagid=1005&tagtoplevel=1005.
4. ANSI Z136.1 (2000) Safe Use of Lasers (Pub 106). Details at http://www.laser-institute.org/onlinestore/index.php3?group=1.
5. ANSI Z136.2 (1997) Safe Use Of Optical Fiber Communication Systems Utilizing Laser Diode and LED Sources (Pub 112). Details at http://www.laserinstitute.org/onlinestore/index.php3?group=1.
6. IEC TC86 Fibre Optics. Details at http://www.tiaonline.org/standards/tag/documents.cfm?tagid=1006&tagtoplevel=1006.
7. Y. Delisle, G. Peng, and B. Lavallee, Designing Safety into Optical Networks, *Laser Focus World,* Aug. 2002.

LASER BANDWIDTH

The transmitter's bandwidth should be twice as large as the transmission data rate.

Discussion

A transmitter (laser and modulator) that is stable at 80 GHz can be used for 40 GHz transmission; likewise, one should have a 200-GHz laser for 100-GHz transmission. As Stellpflug[1] points out,

> To ensure the performance of a 40-GHz optical receiver, the frequency response must be measured to at least 60 GHz (the authors recommend 80). The dominant frequency components are near 20 GHz, while their third harmonics near 60 are essential to reproduce the signal with good fidelity.

Additionally, he asserts,

> Assumptions about the phase response need to be made to extract accurate amplitude information; otherwise, the bandwidth of the optical receiver may appear artificially narrow. Additionally, one needs to know the spectral shape of the laser pulse, and its frequency-bandwidth should be at least ten times larger than the receiver bandwidth.

Reference

1. M. Stellpflug, "Frequency Response Indicates Optical-Receiver Performance," *WDM Solutions*, July 2001, p. 77.

LASER BRIGHTNESS

The brightness of a single-mode laser can be closely estimated by dividing the power-area product by the wavelength squared.

$$B \approx \frac{PA}{\lambda^2}$$

where B = brightness of the beam (watts/steradian)
         P = power of the laser (watts)
         A = area of the radiating aperture
         λ = wavelength

Discussion

This rule derives directly from the concept that the on-axis irradiance of a laser is defined as the brightness divided by the range squared. The logic is as follows.

Brightness is defined as the ratio of the power output to the solid angle into which the beam is projected. So,

$$\text{Brightness} = \frac{\text{Power of the beam}}{\text{Solid angle of the beam}}$$

The solid angle of the beam is the area of the beam at the target, divided by the distance to the target squared.

$$\text{Solid angle} = \frac{\text{area}}{\text{range}^2}$$

The area of the beam is approximately the square of the product of the beam angle, $\approx \lambda/D$, and the range (R).

$$\left(\frac{\lambda}{D}R\right)^2$$

Therefore, brightness is

$$\frac{PD^2}{\lambda^2} = \frac{PA}{\lambda^2}$$

The irradiance that is created also depends on the optical quality of the beam. This is sometimes more fully expressed as[2]

$$\text{Irradiance} = \frac{PA}{\lambda^2 (BQ)^2 R^2}$$

where BQ represents the beam quality, measured as a factor that is equal to unity for diffraction-limited performance and a number exceeding unity for all other cases. Using this formulation, we find that brightness is defined as

$$\frac{PA}{\lambda^2 (BQ)^2}$$

This rule is related to the etendue,[3] which states that $A\Omega \approx \lambda^2$ can be illustrated simply by noting that, for a diffraction-limited beam, the solid angle that is obtained (Ω) is

$$\pi \left(\frac{1.22\lambda}{D} \right)^2$$

The area of the aperture is $(\pi D^2)/4$ so that the product of these two terms is

$$3.67 \, \lambda^2$$

A 1-mW HeNe laser with an aperture of 1 mm will have a brightness of about 2 million W/sr. In contrast, a 1000-W quartz-halogen lamp emits roughly 10 percent of its input power as visible light, over a 4π solid angle. Thus, it has a brightness of about 8 W/sr. Moreover, the laser puts all of its brightness into a very narrow spectral band, while the lamp produces light over a wide range of wavelengths. This means that the spectral brightness of the laser is extremely high.

Additionally, a HeNe laser with a 1-mm beam producing 1 mW radiates as if it were a blackbody of temperature 4.7×10^9 K.[1]

References

1. G. Fowles, *Introduction to Modern Optics*, Dover, New York, p. 223, 1975.
2. G. Golnik, "Directed Energy Systems," in Vol. 8, *Emerging Systems and Technologies*, S. Robinson, Ed., *The Infrared and Electro-Optical Systems Handbook*, J. Accetta and D. Shumaker, Ex. Eds., ERIM, Ann Arbor, MI, and SPIE, Bellingham, WA, p. 451, 1993.
3. A. Siegman, *Lasers*, University Science Books, Mill Valley, CA, p. 672, 1986.

DISTRIBUTED FEEDBACK LASER DRIFT

Typical spectral drift in distributed feedback (DFB) lasers is less than

0. 2 nm/°C
0.01 nm/mA
0.002 nm/°C to case temperature
0.001 nm/year aging drift

Discussion

The transmitter laser must remain within the expected laser wavelength (or frequency) bandwidth and should not drift out of such over time or within various environmental conditions calculated for the expected environments. These specs become increasingly important as DWDM moves to more lasers per fiber.

As mentioned in the reference,[1] the output power should be stable over time, and the side modes should be suppressed to better than 40 dB below the peak output. Moreover, the laser should be optically isolated and should not be affected by spurious reflections from the transmission medium, especially those coming from the first EDFA.

Reference

1. A. Girard et al., *Guide to WDM Technology and Testing,* EXFO Electro-Optical Engineering Inc., Quebec City, Canada, p. 58, 2000.

LED VS. LASER RELIABILITY

Mean time to failures (MTTF) are around 10^6 to 10^7 hr for LEDs operating at 25°C. Conversely, commercially available lasers have an MTTF of about 10^5 hr at 22°C.

Discussion

Advances in materials technology will continue to improve these values but, for now, the above values apply. Also, be aware that the operating temperature of the various lasers has a dramatic impact on the life of these components. Additionally, the mean time between failure (MTBF) of high-tech, high-power lasers is usually several orders of magnitude lower.

These rules are useful for estimating the reliability of various systems and deriving the cost of the systems.

For optical communication, commercially available lasers have failures from defects in the active region, facet damage, and nonradiative recombination in the active region. Much work is progressing toward extending laser reliability.

Reference

1. N. Lewis and M. Miller, "Fiber Optic Systems," in Vol. 6, *Active Electro-Optical Systems*, C. Fox, Ed., of *Infrared and Electro-Optical Systems Handbook*, J. Accetta and D. Shumaker, Ex. Eds., ERIM, Ann Arbor, MI, and SPIE, Bellingham, WA, p. 258–259, 1993.

ON-AXIS INTENSITY OF A BEAM

For a beam with no aberrations, the on-axis intensity, in watts per area, is

$$\frac{PA}{R^2\lambda^2}$$

where R = range, meters
P = beam power, watts
A = transmitting telescope area, meters^2
λ = wavelength, meters

Discussion

If aberrations exist and are characterized by beam quality, BQ, then we get

$$\frac{PA}{(BQ)^2R^2\lambda^2}$$

This form is a direct result of defining beam quality as being a constant that multiplies the beam spread associated with diffraction. As a result of the multiplication, the beam is bigger in each dimension by a factor of BQ, so the area over which the beam is spread is proportional to $1/(BQ)^2$.

This rule is derived from basic laser theory and applies in general. It is limited by the assumption that the beam is propagating without significant atmospheric or other path effects. The discussion below illustrates how that case complicates things.

In many applications, the size of the detector is smaller than the beam at the destination. As a result, the on-axis intensity represents the maximum power that can be delivered into such a detector. Of course, as noted in more detail below, the formula above is the ideal. The presence of aberrations in the beam expander and/or laser, along with atmospheric influences, will reduce the power that can be delivered.

The far-field intensity (I_{ff}) for a circular aperture with reductions due to diffraction, transmission loss and jitter is[1,2]

$$I_{ff} = \frac{I_o TK \exp^{-\sigma^2}}{1 + (1.57\sigma_{jit} D/\lambda)^2}$$

where I_o = intensity at the aperture
T = product of the transmissions of the m optical components in the telescope

K = aperture shape factor found in Holmes and Avizonis;[3] found to be very nearly unity in most cases

k = propagation constant = $(2\pi)/\lambda$

$\Delta\phi$ = wavefront error

σ_{jit} = two-axis rms jitter

D = aperture diameter

λ = wavelength

σ = $k\Delta\phi$, the wavefront error expressed in radians

This leads to a definition of brightness of a laser with jitter and wavefront error[2]

$$\text{Brightness} = \frac{\pi D^2 PTKe^{-\sigma^2}}{4\lambda^2[1 + (1.57\sigma_{jit}D/\lambda)^2]}$$

where P = laser power

We also note that when the wavefront error, σ, and the jitter term are zero, we get the following:

$$\text{Brightness} = \frac{APTK}{\lambda^2} \approx \frac{PA}{\lambda^2}$$

so the on-axis intensity is equal to

$$\frac{\text{Brightness}}{\text{Range}^2}$$

References

1. K. Gilbert et al., "Aerodynamic Effects," in Vol. 2, *Atmospheric Propagation of Radiation*, F. Smith, Ed., *Infrared and Electro-Optical Systems Handbook*, J. Accetta and D. Shumaker, Exec. Eds., ERIM, Ann Arbor, MI, and SPIE, Bellingham, WA, p. 256, 1993.
2. Robert Tyson, *Principles of Adaptive Optics*, Academic Press, San Diego, CA, p. 16, 1991.
3. D. Holmes and P. Avizonis, "Approximate Optical System Model," *Applied Optics* 15(4), p. 1075, 1976.
4. E. Friedman, "On-Axis Irradiance for Obscured Rectangular Apertures" *Applied Optics*, 31(1), Jan. 1, 1993.

OUT-OF-PLANE BEAM

The shape of the lowest-order mode of the out-of-plane beam from a laser diode depends on the waveguide thickness.

Discussion

The equation is

$$w = d_g\left(0.321 + \frac{2.1}{V^{3/2}} + \frac{4}{V^6}\right) \text{ for } 1.8 < V < 6$$

Variable w refers to the beam shape parameter used in a Gaussian description. That is, the beam is described as

$$\exp\left(\frac{-x^2}{w^2}\right)$$

Here, d_g is the waveguide thickness, which is about 560 nm for a laser that produces light at 1300 nm. V is related to d_g by the following equation:

$$V = d_g k\sqrt{n_g^2 - n_c^2}$$

where g refers to the properties of the guide, and c refers to the properties of the cladding. Often $n_g - n_c$ is about 0.2. As usual, k is the propagation constant $2\pi/\lambda$. For the far-field beam size,

$$\theta_{ff} = \tan^{-1}\left(\frac{\lambda}{\pi w_o}\right)$$

where the o subscript refers to the initial size of the beam and we have a slightly different equation. In this case,

$$w = d_g\left(0.31 + \frac{3.15}{V^{3/2}} + \frac{2}{V^6}\right) \text{ for } 1.5 < V < 6$$

of w_o is used in the equation.

Reference

1. E. Garmire, "Sources, Modulators, and Detectors for Fiber-Optic Communication Systems," Chap. 4 in *Fiber Optics Handbook*, M. Bass, Ed. in Chief, and E. Van Stryland, Assoc. Ed., McGraw-Hill, New York, p. 4.13, 2001.

LASER DIODE RELIABILITY AND OPERATING TEMPERATURE

If the laser diode's operating temperature is reduced by about 10°, its lifetime will double.

Discussion

It is widely known that optical power output increases as a laser diode's temperature decreases. A concern arises in situations where the diode is expected to deliver a constant power mode. Without temperature control, the diode can change modes and wavelengths. Without proper cooling, as might happen if the heat sink is not adequate, the power circuit will drive the diode harder and harder, attempting to make up for the lost efficiency at the higher temperature. This situation can lead to a runaway and destruction of the diode, and mode hops and changes in wavelength will still occur. A different failure mode, but perhaps even more expensive, is the gradual degradation of performance over time, requiring multiple service calls.

References 2 and 4 also indicate that the wavelength of a typical GaAlAs diode will increase on the order of 0.25 to 0.27 nm for a 1°C rise in temperature. Finally, we must point out that extending the life of the diode itself is not a system solution, since other items, like the cooler, may then determine the overall life.

References

1. P. Hobbs, *Building Electro-Optical Systems, Making It All Work,* John Wiley & Sons, New York, p. 368, 2000.
2. http://optima-prec.com/notes.htm, 2002.
3. Staff of Profile Optische Systeme GmbH, "Basic Notes Laser Diodes," http://www.timesfiber.com/techpdf/1006a-tn.pdf, 2002.
4. Staff of JDS Uniphase, SDL-6300 Series, 2 to 4 W, 920 and 980 nm High-Brightness Laser Diodes, http://www.psplc.com/downloads/acrobat/jdsu/sdl/JD6300laserdiode.pdf, 2002.

THERMAL CONTROL IN TRANSMITTERS

A typical dense wavelength division multiplexing (DWDM) module changes wavelength with temperature with a sensitivity of about 0.2 to 0.6 nm/°C.

Discussion

All of the components of optical telecom systems are susceptible to performance change with temperature. The use of temperature sensors in thermal management control loops is an important part of all system designs.

Figure 13.1 is an example of the temperature sensitivity of a critical component in any fiber system. It shows how the operating wavelength of a DWDM laser module changes wavelength with temperature. Obviously, to avoid crosstalk between channels, significant drift in operating wavelength cannot be tolerated.

At least for the near future, electronic management of temperature will be the chosen method of managing this type of problem. At some future date, systems may exist that are insensitive to temperature variation or that have inherent self-correcting optical components. Until that time, it is important for the system designer to understand the available options for managing this type of problem.

Control can be implemented by using coolers that are controlled by a computer that integrates the data from temperature sensors. Another option is to determine an appropriate operating temperature and include computer-controlled heaters that assure an appropriate input that stabilizes the system component operating temperatures.

In mentioning computer implementation of control loops for temperature stabilization, we are not ignoring analog computers. They offer a simple and low-cost mechanism for controlling the components that heat or cool the optic systems in a network.

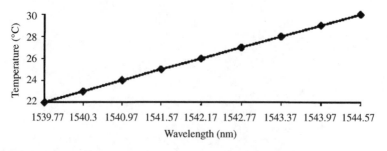

Figure 13.1   Temperature dependence of wavelength for a typical DWDM laser module.

For example, a 25-GHz laser line spacing is about 0.2 nm. Therefore, according to this rule, the laser temperature must be controlled to a small fraction of a degree Celsius over the entire environmental temperature range.

Reference

1. C Patrick Lyons, "Recent Developments for Optical Component Temperature Sensing," *Proceedings of the National Fiber Optic Engineers Conference,* 2001.

FDBR SPECTRAL RESOLUTION

Using a fiber grating distributed Bragg reflector (FDBR), a spectral resolution of 0.1 nm can be achieved.

Discussion

A number of technologies must come into maturity to assure the very high bandwidth achieved by DWDM. A number of concepts exist for generating the narrow wavelength carriers that enable this type of high-performance fiber system (see Figs. 13.2 and 13.3). Among these is the photorefractive Bragg grating in glass fibers. These are commonly called *hybrid fiber grating distributed Bragg reflector* lasers. By exploiting a single mode of laser cavity, a spectral bandwidth of

$$\Delta\lambda = \frac{\lambda^2}{2\{n_g L_g + n_s L_s + n_f[\tanh(\kappa L_{\mathrm{Bragg}})/2\kappa]\}}$$

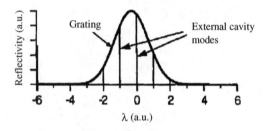

Figure 13.2   Reflectivity vs. wavelength (a.u. = arbitrary units.

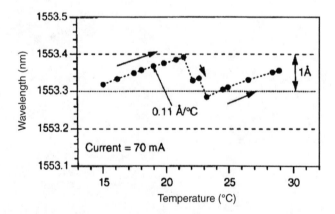

Figure 13.3   Wavelength tuning of a fiber DBR vs. temperature.

is achieved. In the equation, the subscript g refers to the index (n_g) and length (L_g) of the gain region, and s describes the spacer between the grating and gain chip. Variable n_f is the fiber index, κ is the grating coupling constant, and L_{Bragg} is the Bragg grating length. Free spectral range is on the order of 1 Å for a short cavity.

These devices have an interesting temperature property. Due to the fact that the change in the index of the semiconductor gain chip versus temperature is larger than that of the glass fiber, the Fabry–Perot modes will shift faster than the Bragg reflectivity envelope. Eventually, whichever mode is being transmitted will move from under the grating reflectivity curve and jump to a neighboring mode. Temperature stabilization devices and circuits are useful for avoiding this problem.

Reference

1. M. Ziari et al., "Fiber-Grating Based Dense WDM Transmitters," *Proceedings of the National Fiber Optic Engineers Conference*, 1997.

SPECIFYING LASER TRANSMITTERS

Use common sense in specifying laser transmitters.

Discussion

Consider the following rules:

1. Avoid open packages, as they invite mechanical and environmental damage.

2. Consider active temperature control. An increase of 10 K has a devastating impact on diode reliability. As much as 50 percent of its life might be lost by such an increase.

3. Make sure the electronic drivers for the diode are fully equipped to protect the laser. This is a specialized area, so choose your vendor carefully.

4. Operate the laser with plenty of margin. This precaution can greatly extend device life.

5. Use standard wavelengths to control expenses related to having vendors create custom items.

6. Use standard packages. As above, be wary of asking for custom components that add lots of cost and extend the schedule.

7. Don't specify characteristics that are hard to control. This is another cost risk. Work with your suppliers to be assured that they have catalog items that meet your needs before having them invent something new.

8. Consider active alignment for components whose positions are critical. Even though more complex, a positioning technology working from an error signal in the system can guarantee system success, especially as tolerances grow more critical.

9. Don't overdo the specification of optical performance. Use full width, half maximum or some similar criterion rather than a specific beam shape.

Reference

1. L. Hillis, "Rules of Thumb: Help in Specifying Laser Diodes," *Photonics Spectra*, June 2001, p. 167.

VERTICAL CAVITY SURFACE EMITTING LASER GAIN

The per-pass gain of a vertical cavity surface emitting laser (VCSEL) is about 1 percent, which is approximately that of a HeNe laser.

Discussion

A VCSEL has an efficiency of about 1 percent. This seems low until one realizes that this is about the standard for a HeNe laser.

LASER LINEWIDTH MEASUREMENT

Measure the linewidth at –20 dB below the peak and divide by 10.

Discussion

DWDM lasers have narrow peaks that are very close to each other. Heterodyning techniques can be used to accurately measure the linewidths. A reference laser is mixed with an unknown source and applied to an optoelectric converter (typically a high-speed photodiode) and analyzed by a microwave spectrum analyzer. The photodiode's output represents the cross correlation of the signals from the two lasers in the frequency domain. The mixed signal has a Lorentzian line shape. One could measure the linewidth at the –3-dB level on the heterodyne signal, but noise will often interfere with an accurate measurement. If the signal is measured, the Lorentzian peak is at –20 dB below the peak. The linewidth at this point is ten times the actual linewidth. If measured at –40 dB, it is 100 times the actual linewidth.

Reference

1. B. Nosratieh, "Characterize Lasers for DMDM Transmission," *Test and Measurement World*, Jan. 2001, pp. 25–32.

SINGLE LONGITUDINAL MODE LINEWIDTH

The linewidth of single longitudinal mode laser is about 0.1 nm.

Discussion

This is true if no special stabilization is used. At 1550 nm, this is a frequency bandwidth of about 12.5 GHz.

Reference

1. I. Jacobs, "Optical Fiber Communication Technology and System Overview," Chap. 2, *Fiber Optics Handbook*, M. Bass, Ed., in Chief, and E. Van Stryland, Assoc. Ed., McGraw-Hill, New York, p. 2.6, 2001.

OPTICAL DECIBELS

$$dBm_{opt} = 10 \log(P_{opt})$$

where dBm_{opt} = optical decibels
P_{opt} = power in milliwatts

Discussion

Typically, an optical spectrum analyzer measures power in units of dBm or decibels in relationship to 1 mW (not 1 W). Telecommunication lasers tend to have peak powers in the range of 10 to 100 mW, or 10 to 20 dBm, with a 30-dBm separation between the main mode and side modes. Also, please note that most spectral test equipment normalizes the measurement to a 0.1-nm bandwidth, which is getting unusually coarse as DWDM packs more and more wavelengths into a given band.

The above equation allows users to quickly convert between optical decibels and linear power, which is useful in determining signal-to-noise parameters. The reader should take note that the common usage of decibels in electrical applications (as opposed to optical) uses a multiplier of 20 rather than of 10.

References

1. B. Nosratieh, "Characterize Lasers for DMDM Transmission," *Test and Measurement World,* Jan. 2001, pp. 25–32.
2. http://www.udtinstruments.com/pdf/FO-Tutorial.pdf, 2002.

14

Wavelength Selection

There are strong economic motivations to increase the number of modulated wavelength channels that can be accommodated by a fiber-based telecom system. Not only does the designer avoid the cost of adding new cable, but it is possible to dramatically increase the bandwidth per existing channel.

This requires, of course, the creation of new hardware and system management tools that can squeeze more optical channels into each fiber and then find them at the other end, separate them, and deal with them individually. This implies constantly improving the capabilities of filters, lasers, amplifiers, multiplexers, and other components that must "handle the photons with care." From the system management point of view, standards must evolve along with the hardware so that transmission penalties are minimized, bandwidth is maintained, and growth paths are not shut off.

As a result of the press of technologies that support the concentration and ultimate division of wavelengths, we have included this chapter to specifically address wavelength separation. It addresses antiresonant reflecting optical waveguides (ARROWs), waveguides, gratings, multiplex and demultiplex components, filters of various types, and so on. Throughout, we have tried to include practical issues, such as controlling the performance of these components in the real environments where they will be used, with particular attention to the thermal environments that are encountered in the field. We have included most of the popular filter systems, including those that use fiber Bragg gratings, conventional gratings, and waveguide methods.

Gratings, particularly those created within fibers, such as fiber Bragg gratings, get a lot of attention, with additional information on Fabry–Perot etalon technology. Some of the features of Mach–Zehnder interferometers and their role in multiplexing are also addressed. As noted above, temperature management and wavelength stabilization of these devices is critical to their implementation and receives some emphasis. For those who manufacture such devices, some detail is provided on UV photoexposure of fibers, including those pretreated with hydrogen and Ge doping. The Arhennius equation (also described in other chapters; see the index) comes up from time to time as a model for accelerated testing of devices that will be operated at elevated temperatures.

Throughout the chapter, we have concentrated on rules that put performance into quantitative terms so that the readers can do calculations to estimate how the components will perform in their systems.

Finally, we include a table that summarizes some aspects of the performance of key filter technologies as a function of their DWDM spacing.

ACOUSTO-OPTIC TUNABLE FILTER BANDPASS

The bandpass of an acousto-optical tunable filter (AOTF) (in microns) can be approximated by

$$\Delta\lambda \approx \frac{0.8\lambda^2}{L\Delta n}$$

where $\Delta\lambda$ = filter bandpass
     λ = center wavelength
     L = transducer width
     Δn = crystal's bifringence at wavelength λ

and the wavelength passed is

$$\lambda = \frac{\Delta n V}{f}$$

where V = acoustic velocity in the crystal
     f = applied acoustic pressure frequency

Discussion

AOTFs have been used since the 1980s to tune the wavelengths for imaging and telecommunication applications, as they can be very narrow and very fast. These filters are exotic crystals with an index of refraction that is a strong function of strain. Thus, a substantial change in the index of refraction can occur for a very small strain. An induced pressure standing wave can cause a periodic density variation, resulting in a "Bragg" effect. A piezo-actuator can be used to pump an acoustic standing wave (at an RF frequency) into the crystal, causing rapid and accurate transmission wavelength selection. Often, these crystals are polarized, so care should be taken to assure proper polarization coupling.

Note that, at a wavelength of 1.55 μ, bandwidth is approximated by

$$2/L\Delta n$$

Reference

1. S. Kartalopoulos, *Introduction to DWDM Technology*, IEEE Press, Piscataway, NJ, p. 82, 2000.

ARRAYED-WAVEGUIDE GRATINGS

The referenced works[1,2] encourage the use of parabolic horns to assist in separating the closely spaced optical signals. They point out that the profile of the parabolic horn, $W(z)$, can be expressed in the following way:

$$W(z) = \sqrt{2 a \lambda_g + D^2}$$

where α = a constant less than unity
λ_g = wavelength in the waveguide
D = core width

By appropriate selection of parameters, a double-peaked beam can be obtained.

Discussion

The attractiveness and efficiency of arrayed-waveguide (AWG) gratings (wavelength division multiplexing) does not come without some technical challenges. Clearly, one of the key concerns is management of wavelength channels, which can get quite narrow. Arrayed-waveguide gratings have become popular for spectral separation of WDM signals.

The application of AWGs is widespread because they address the problem of demultiplexing many wavelengths that might be carried within a single fiber. Any device used for demux must also perform this function with low loss, ease of use, and low crosstalk. AWGs also have these features. At the same time, their performance within the narrow spectral range exiting the fiber demands that their temperature be managed carefully and that other factors, such as stress-induced birefringence, be taken into account in their design and in the design of systems that make use of them. The references discuss these issues and offer a proposed solution for flattening the spectral bandwidth of the grating, thereby relaxing the demands on its temperature control and allowing for additional tolerance in the wavelengths of the transmitted beams that exit the fiber.

References

1. K. Schmidt et al., "Wavelength Flattened Arrayed-Waveguide Grating Multiplexer with Low Polarization Sensitivity," *Proceedings of the National Fiber Optic Engineers Conference,* 1997.
2. C. Peucheret and C. Herben, "First Design of Band-Flattened GHz Silica PHASAR," Research Center COM, T.U. Denmark, and Delft University of Technology, 2002, available at http://www.ecitele.com/meteor/D11.pdf.

Antiresonant Reflecting Optical Waveguide Antiresonance Thickness

A waveguide can be formed by a layer of materials that mimic the operation of a Fabry–Perot etalon except that, in this case, we want the thickness of the layers to be such that radiative loss is low. The following equation estimates the thickness of the thin cladding (middle) layer in a stack of three layers that meets this requirement.

$$t \approx \frac{\lambda}{4n_2}(2N+1)\left(1 - \frac{n_1^2}{n_2^2} + \frac{\lambda^2}{4n_2^2 d_1^2}\right)^{-1/2}$$

where $N = 0, 1, 2...$

In this case, the top layer of silicon dioxide is the "fiber." Below it is the thin polysilicon layer. Below the poly layer is another layer of SiO_2. The index of the top layer is n_1, and n_2 is the index of the bottom layer. The thickness of the fiber layer (which is on top) is d_1.

Discussion

Here, the thin central layer is polysilicon, while the layers above and below it are SiO_2. The reader will note that the above expression depends only on the thickness of the top layer of SiO_2. In practice, this expression is usually equal to odd multiples of $\lambda/4n_2$. The thickness of the bottom layer maintains the minimum loss condition when it is odd multiples of $d_1/2$.

Reflectivity at boundaries of different indices of refraction is a widely known phenomenon. Think of the reflected image that can be seen as you look outward through a window at night. It is also true that the thickness of a layer controls the amount of reflection that occurs. The reference shows the thickness of a layer that cancels the reflection. The result is generally applicable, although the reference was developed to calculate the "antireflectance" thickness for the top SiO_2 layer.

An antiresonant reflecting optical waveguide (ARROW), depends on a Fabry–Perot-like resonator, so-called *antireflection*, as opposed to total internal reflection. As the reference points out, for ARROW structures, the bottom silica layer often has a thickness of about 2 microns, while the top layer has a thickness of approximately 4 microns. Thus, the ARROW architecture exploits a silica-like fiber core and the ensuing interaction with a different index cladding. Additionally, TM polarization modes have much more attenuation than TE modes, so devices constructed in this manner are very polarization sensitive.

There is a rule in Chap. 3 concerning the chip yield for various photo-lithographic processes. It also applies to the manufacturer of ARROW devices.

References

1. C. Gorecki, "Optical Waveguides and Silicon-Based Micromachined Architectures," in P. Rai-Choudhury, Ed., *MEMS and MOEMS Technology And Applications,* SPIE, Bellingham, WA, pp. 217–218, 2000.
2. M. Duguay et al., "Antiresonance Reflecting Optical Waveguides in SiO_2-Si Multilayer Structures," *Applied Physics Letters* 49(13), July 7, 1986.
3. U.S. patent 5,692,076, "Antiresonant Waveguide Apparatus for Periodically Selecting a Series of at Least One Optical Wavelength from an Incoming Light Signal," V. Delisle et al., 1995.

ARRAYED-WAVEGUIDE GRATING STABILITY

The wavelength shift of an AWG varies with temperature and time and has a typical form of

$$\Delta\lambda = A\left(1 - e^{-Bt}\right)$$

where $\Delta\lambda$ = shift in the center wavelength of the filter
   t = aging time
   A = center wavelength shift for an infinite exposure time
   B = a parameter that defines the rate of change of the center wavelength

Discussion

Accelerated testing of arrayed-waveguide grating modules with waveguide-fiber fusion spliced connections offers some insight into how integrated components are degraded by temperature.

Takasugi et al.[1] exposed such devices to elevated temperatures ranging from 100 to 180°C. This allows reliable estimates of the shift of the center wavelength at ~80°C. The parameter, A, can be estimated from the Arrhenius model, which has the form of

$$A = Ce^{-E/kT}$$

where C = a constant measured at normal operation conditions
   k = Boltzmann's constant
   T = elevated temperature used during the accelerated tests
   E = activation energy for this particular process

This model for accelerated testing appears elsewhere in this book.

Reference

1. S. Takasugi et el., "Reliability of Arrayed Waveguide Grating Modules with Waveguide-Fiber Fusion Spliced Connection," *Proceedings of the National Fiber Optic Engineers Conference*, 1998.

EXTRACTION ANGLE FOR BLAZED BRAGG GRATINGS

The Bragg equation defines the angle (θ) at which light is extracted from the core of a fiber for a wavelength λ and blaze angle θ_{blaze}. The equation is shown below.

$$\lambda = \frac{n\Lambda}{\cos\theta_{blaze}}(1 + \cos\theta)$$

where n = index of refraction of the core
Λ = unblazed grating periodicity

Discussion

Bragg gratings are widely used to extract light from fibers. Most implementations do not have a *blaze* angle, but it is possible to incorporate a blaze. The angle at which the light leaves the fiber depends on the blaze angle (ϑ_{blaze}), as is true in all grating applications. To achieve low polarization effect, this angle should be less than 20°.

As is true of all gratings, the dispersed light has a different angle of emergence from the fiber for each wavelength of interest. This feature can be exploited by *chirping* the grating. That is, the periodicity of the grating changes along its length so that light of a single wavelength finds its way to a focal point, regardless of where along the grating it emerges. To achieve this effect, the frequency of the grating must be consistent with the following equation. The focal length of the grating, f, is approximated by

$$f \approx \frac{\Lambda^2 n\sin\theta}{\lambda\cos\theta_{blaze}C_{chirp}}$$

C_{chirp} = grating period change per unit length

An addition benefit of chirping the grating is that it provides dispersion compensation when employed with an appropriate dispersion-compensating fiber. As mentioned elsewhere in this chapter, Bragg gratings can be mechanically stressed to achieve wavelength tuning or polarization management.

References

1. C. Koeppen et al., "High Resolution Fiber Grating Optical Network Monitor," *Proceedings of the National Fiber Optic Engineers Conference*, 1998.
2. A. Benito et al., "Chirped Fiber Grating-Based Fiber Optic Communication Evaluator," *Optical Engineering*, 38(10), Oct. 1999.
3. http://www.ofsoptics.com/products_info, 2002.

BRAGG WAVELENGTH DEFINED

The Bragg wavelength is given by

$$\lambda_b = 2n_{eff}\Lambda = \lambda_{uv}/[2\sin(\alpha)]$$

where λ_b = Bragg wavelength
   n_{eff} = effective index of refraction
   Λ = period of the index of refraction variation
   λ_{uv} = wavelength of the UV light used to induce the Bragg grating
   α = angle between the interfering UV beams

Discussion

A Bragg grating can be created internal to a GeO_2-doped fiber by the interference of two UV lasers. The strong (>10 W/cm^2) ultraviolet (UV) interference pattern created in the fiber causes a spatially periodic change of the index of refraction. This effect is complex due to a combination of stress effects including heating, formation of defect zones, and electronic excitations in the molecular structures. With proper annealing, the UV light imposes a semi-permanent change in the index of refraction as a result of the photosensitivity of the fiber material. Even visible light can induce this effect. Exposures of a few minutes at 100 to 1000 $mJ/cm^2/$ pulse are used. When even higher radiation doses are used, nonlinear photosensitivity is observed that allows the grating effect to be imprinted into the fiber in a single pulse. Moreover, hydrogen doping of the fibers prior to exposure can enhance the effect. Subsequent heating of the fiber drives out the unbound hydrogen. The hydrogen in the strongest parts of the exposure light forms the species responsible for the index of refraction changes.

Othonos et al.[3] point out that, since the first observation of photoinduced gratings in germanium doped optical fibers, there have been many important advances in Bragg grating technology. Their acceptance as standard devices in the designer's arsenal has depended on methods for ease of production. In addition, these devices can be installed anywhere within the fiber and have the wavelength breadth to work well with EDFAs. They also point out that these very simple devices are an important component of DWDM systems. Ease of use is a factor as well, since they are just a piece of fiber.

References

1. A. White and S. Grubb, "Optical Fiber Components and Devices," in *Optical Fiber Telecommunications*, I. Kaminov and T. Koch, Eds., Academic Press, San Diego, CA, pp. 273–274, 1997.

2. C. Gorecki, "Optical Waveguides and Silicon-Based Micromachined Architectures," *MEMS and MOEMS Technology and Applications*, in P. Rai-Choudhury, Ed., SPIE, Bellingham, WA, pp. 284–287, 2000.
3. A. Othonos, X. Lee, and D. Tsai, "Spectrally Broadband Bragg Grating Mirror for an Erbium-Doped Fiber Laser," *Optical Engineering* 35(4), pp. 1088–1092, April 1996.

DECAY OF BRAGG FILTER

The decay of the UV-induced index change can be estimated from

$$\eta = \frac{1}{1 + \exp\left(\dfrac{E_d - \Delta E}{kT}\right)}$$

where
η = UV-induced change in index of refraction
ΔE = energy spacing between levels
E_d = detrap energy
k = Boltzmann's constant
T = a temperature equal to approximately 5200 K

Discussion

Strong UV light directed into a fiber can cause color centers and traps, resulting in a variation in the index of refraction. This can be used to deposit a Bragg grating into a fiber, especially when the fiber is doped with various materials (often Ge). The permanence of UV-induced Bragg filters is a concern for the designer and user, as these traps can disappear with time and temperature. The above relationship allows one to estimate the fraction of the UV-induced index change that fades due to annealing from temperature or time.

Traps with activation energies less than $E_d = kT \ln(\nu_o t)$ are thermally depopulated, and those with greater energy are filled. Note that this is similar in form to the Arrhenius equation found elsewhere in this book. Some types of fiber Bragg gratings (FBGs) are more persistent than others.

Decay of a UV-induced index change versus time has an initial drop and then levels off to a slow decay.

Also, using the above relationships, one can eliminate the higher-energy traps with a thermal anneal. Annealing is important for long-term stability, as a 1-hr anneal at 100°C results in only 5 percent decay at room temperature over 100 yr.

Reference

1. A. White and S. Grubb, "Optical Fiber Components and Devices," *Optical Fiber Telecommunications*, I. Kaminov and T. Koch, Eds., Academic Press, San Diego, CA, pp. 285–287, 1997.

DENSE WAVELENGTH DIVISION MULTIPLEXING DEVICES

Here, we compare four important types of dense wavelength division multiplexing (DWDM) multiplexers and demultiplexers.

They are arrayed wavelength gratings (AWGs), fiber Bragg gratings (FBGs), dielectric bandpass filters, and hybrids of FBG and dielectric coated thin-film bandpass WDM filters.

DWDM device	Advantages	Disadvantages
AWG DWDMs	• Good λ control and λ stability • Low insertion loss • Low cost, cost does not increase with increase of channel count	• Narrow useful bandwidth @ –0.5 dB • PDL and polarization dependent λ • High crosstalk to all other channels • AWG–fiber interface reliability
FBG-based DWDMs	• Broad flat passband width @ –0.5 dB • High channel isolation • good λ control and λ stability • Low PDL • Low passband ripple	• High cost using multiport circulator • T-control or T-stabilized package • High insertion loss • Cost proportional to channel count
Thin-film filter-based DWDMs	• λ stability for ion-assistant coating • Low insertion loss and small PDL • Broad bandwidth @ –0.5 dB • Small temperature sensitivity	• Add channel proportional to add cost • Small figure of merit • Low yield of qualified filters • High ripples in passband
Hybrid DWDMs (FBG/thin film filters)	• Substantially relaxed filter requirements • λ accuracy and stability • Small PDL • Broad passband width @ –0.5 dB • High channel isolation • Cost-effectiveness	• Add channel proportional to add cost • Relatively higher insertion loss • T-compensated or T-control package

Discussion

Components able to manage the many (and narrow) wavelengths used in DWDM have advanced dramatically in capability in the last decade or so. The components mentioned in the table summarize the state of the art at the time of publication of this book; more advances can be expected as new materials and production methods are put to work. Indeed, one need only look at the collection of current journals to see the advance presentation of the new devices. Recent history shows that the time to progress from first demonstration to use in actual system applications is short and getting shorter. In addition, new advances in MEMS and nanotechnology can be expected to add to the arsenal available to the system designer. For example, not included here but in a similar category are the many new ideas emerging for optical switching.

Reference

1. J. Pan and Y. Shi, "0.4 nm Channel Spacing Hybrid Dense WDM Multiplexer and Demultiplexer Using FBGs and Thin-Film Filters," *Proceedings of the National Fiber Optic Engineers Conference,* 1988.

FABRY–PEROT ETALONS

The transmittance and bandwidth of a Fabry–Perot interference filter can be represented by

$$\text{Peak transmittance} = \frac{1}{(1 + A/T)^2}$$

and

$$\text{Bandwidth} \approx \frac{(1 - R)}{(\pi m R^{0.5})}$$

where A = absorptance
R = reflectance
T = transmittance
m = order of interference of a semi-transparent aluminum film

Discussion

This rule is good for explaining and understanding the effects of layer absorptance and reflectance in filter design for cases in which the finesse (resolution) of the filter is low.

These filters usually consist of a thin pair of highly reflective semi-transparent thin films separated by a thin layer of dielectric. Transmission can be increased by thinner deposition of the metal films; however, the bandwidth tends to become greater. As DWDM forces more wavelengths closer together, these components are finding use in the network elements of optical telecom.

The reader may find that the first equation above is more effectively expressed as the exact equation,

$$\text{Peak transmittance} = \left[1 - \frac{A}{(1 - R)} \right]^2$$

This equation becomes the one shown in the rule when one notes that $T + A + R = 1$.

The reader is cautioned that high-finesse Fabry–Perot gratings may be limited by mirror flatness. Finesse is inversely proportional to wavefront error. Thus, a Fabry–Perot (FP) using hundredth wavelength mirrors will have a finesse of only 50.

WAVELENGTH AS A FUNCTION OF STRAIN IN FIBER BRAGG GRATINGS

The change in wavelength with strain in a fiber Bragg grating is linear.

Discussion

A typical relationship between wavelength and strain is given below.

$$\lambda = 1550.085 + 1.191 \times 10^{-3} \, S$$

where λ = wavelength in nanometers, which is the wavelength in air
   S = strain in dimensionless units of microstrain—the ratio of the change in length to the initial length

The linear relationship is the foundation for using these gratings as strain gauges. At the same time, the designer must beware so temperature changes do not create fictitious results. The relationship of wavelength to strain is surprisingly linear, as indicated by any number of test data. Fiber Bragg gratings are described in rules elsewhere in this chapter. Not only do they provide an exquisitely narrow bandpass, they can be used as tunable devices. By merely pulling or compressing the fiber, the period of the fiber Bragg grating changes, thus changing the wavelength as described by this rule.

The above equation was developed for, and the data taken at, the C band, but the concept holds in other optical telecom bands, albeit with potentially different coefficients.

Of course, the above also shows that the shift can be modeled by 1.19 picometers of wavelength per microstrain in this region of bandpass. Or,

$$\Delta\lambda = 1.191 \times 10^{-3} \, S$$

Another advantage of these components is that they exhibit linear behavior at wavelengths consistent with the performance of EDFAs. This is attractive, as it allows components of this type to be embedded in fibers to fully exploit the compact features of these devices. In addition, the authors of Ref. 1 found that the bandwidth and reflectivity of the grating were not affected as the strain was introduced. This bodes well for devising tunable fiber components that work within the bandpass of EDFAs

Reference

1. A. Othonos, X. Lee, and D. Tsai, "Spectrally Broadband Bragg Grating Mirror for an Erbium-Doped Fiber Laser," *Optical Engineering* 35(4), p. 1091, April 1996.

FIBER BRAGG TEMPERATURE DEPENDENCE

A typical temperature-dependent dispersion of -0.011 nm/°C is common in fiber Bragg gratings.

Discussion

Fiber Bragg gratings are an important filter technology for fiber systems. They are attractive because they are in the fiber, reliable, and small. As in any optical filter, FBG must have good definition of its transmission wavelength and strong suppression of harmonics (sidelobes). These performance features are particularly important for WDM systems, where channel spacings are measured in units of 0.1 nm.

Any number of references point out that the wavelength of a FBG is expressed as

$$\lambda = 2\, n_{eff} \Lambda$$

where n_{eff} is the effective refractive index of the FBG. Λ is the period of the refractive index. See, for example Refs. 1 and 4. The thermal expansion coefficient is expressed as

$$\alpha_{eff} = \frac{d\lambda}{\lambda\, dT} = \frac{d\Lambda}{\Lambda\, dT} + \frac{1}{n}\frac{dn}{dT} = a_f + \frac{1}{n}\frac{dn}{dT}$$

where α_f is the thermal expansion coefficient of the fiber core. From this, the references calculate an α_{eff} of 7.1×10^{-6} per °C, which is a wavelength temperature coefficient of 0.011 nm/°C at 1550 nm, as noted above. The equation above is general, but the specific results apply to a particular fiber used in the reference authors' experiments.

Reid and Ozcan[2] find a temperature coefficient of 0.0088 nm/°C for another particular fiber. This reference also includes information on the cryogenic performance of fibers that the reader might find useful.

References

1. J. Pan, Y. Shi and S. Li, "Zero-Wavelength Shift Temperature Stabilized Multi-FBG Package," *Proceedings of the National Fiber Optic Engineers Conference*, 1998.
2. M. Reid and M. Ozcan, "Temperature Dependence of Fiber Optic Bragg Gratings at Low Temperatures," *Optical Engineering*, 37(11), p. 237, Jan. 1998.
3. Highwave Optical Technologies reports values of less than 12 pm/°C. Catalog and specification sheet are available at http://www.laser2000.co.uk/pdfs/view2.pdf.
4. O. Frazão, et al., "Optical Fibre Embedded in a Composite Laminated with Applications to Sensing," *Proceedings of Bianisotropics 8th International Conference on Electromagnetics of Complex Media*, Sept. 27–29, Lisbon, Portugal, 2000.

FIBER GRATING CREATION

The bandwidth (full width, half maximum) achieved by fiber Bragg gratings can be expressed as[1,2]

$$\frac{\Delta\lambda}{\lambda_B} = R\sqrt{\left(\frac{n_1}{2n_o}\right)^2 + \left(\frac{1}{N}\right)^2}$$

where n_0 = average refractive index in the core of the fiber, and s varies
        from 0.5 to 1.0 for reasons described in the "Discussion" section

   λ_B = operating wavelength of the grating
   $\Delta\lambda$ = achieved bandwidth
   N = number of grating planes
   n_1 = strength of the modulation of the index of refraction

Discussion

The manufacture of fiber gratings has undergone great advances in recent years. One technology of particular interest is the ability to impose grating features in fibers by doping with GeO and then exposing the fiber to UV radiation. Furthermore, initial treatment with hydrogen enhances the UV sensitivity of the fiber. After exposure to hydrogen at elevated temperature and pressure, an exposure time is required to create a permanent change in the index of refraction.

   The parameter λ_B is the Bragg wavelength defined by

$$\lambda_B = 2n\Lambda$$

where n = effective index of refraction of the grating material
        Λ = grating pitch

   Another factor of interest is the reflectivity of the grating, which is defined as follows:

$$R = \tanh^2(KL)$$

where L is the length of the grating in the fiber, and K is defined as

$$K = \pi n_1/\lambda_B$$

K is often called the *coupling coefficient* of the grating.[4] The larger the KL, the larger the reflectivity of the grating. For example, for $KL = 1$, $R = 58$ percent. When KL is greater than 1, the reflectivity of the grating is called *strong*, while, for $KL < 1$, it is called *weak*.

   Note that a slightly different form of the equation is found in Refs. 2 and 3.

References

1. P. Russell, J. Archambault, and L. Reekie, "Fibre Gratings," *Physics World*, Oct. 1993.
2. S. Pitassi, F. Pozzi, A. Zucchinali, "Optical Fiber Grating: Narrow Band Transmission Filter Fabrication and Packaging Problems," *Proceedings of the National Fiber Optic Engineers Conference*, 1996.
3. S. Kartalopoulos, *Introduction to DWDM Technology*, IEEE Press, Piscataway, NJ, p. 78, 1999.
4. K. Hill, "Fiber Bragg Grating," Chap. 9 in *Fiber Optics Handbook*, M. Bass, Ed. in Chief, and E. Van Stryland, Assoc. Ed., McGraw-Hill, New York, p. 9.4, 2001.
5. http://www.academicpress.com/optics/appsnonlinear.pdf.

GRATING BLOCKERS

If you keep the spectral range to less than 90 percent of an octave, you don't need spectral blockers.

Discussion

This rule applies to diffraction gratings but, generally, not etalons, as they operate in higher multiple orders. Generally, a grating will diffract light of a wavelength λ at the same angle as light of the wavelength $\lambda/2$, $\lambda/3$, ... and so on. However, there is always some leakage and uncertainties, so it is wise to reduce this by another 10 percent; hence the references suggest the 90 percent of the above rule.

For diffraction gratings, the well known grating equation[1]

$$m\lambda = d(\sin\alpha + \sin\beta)$$

is satisfied whenever m is an integer. Alpha and beta are the incident and diffracted angles from normal, respectively. Unfortunately, this means that light of multiple (where m is an integer) colors will overlap and fall on the same angle. This can be eliminated by pesky and costly order-blocking filters or by adhering to the above rule.

The range of wavelength (F_λ) for which this superposition of other orders does not occur is called the *free spectral range* and can be calculated[1] by assuming that, in the order m, the wavelength that diffracts along the direction λ_1 in order $m + 1$ is $\lambda_1 + \Delta\lambda$, where

$$\lambda + \Delta\lambda = \frac{m+1}{m}\lambda_1$$

and then, applying the 90 percent suggestion,

$$F = 0.9\Delta\lambda = \frac{0.9\lambda_1}{m}$$

For example, if you need to split a spectrum, you can go from about 2 to 1.1 microns without a blocker, or 1.5 to 0.82 microns.

References

1. C. Palmer, *Diffraction Grating Handbook*, Thermo RGL, Rochester, NY, pp. 14–17, 2000.
2. http://www.thermorgl.com/library/handbook4/chapter2.asp.

GRATING EFFICIENCY AS A FUNCTION OF ANGLE

A typical grating's efficiency is a strong function of polarization, blaze angle, and diffraction angle. Generally, one can assume a 50 percent or higher grating efficiency if the diffraction angle lies between

$$\beta \approx 1.1\theta + 2$$

and

$$\beta \approx 0.56\theta + 0.3$$

where β = diffracted angle (in degrees)
        θ = blaze angle (in degrees)

Also, one can assume a 20 percent efficiency throughout a larger range of diffraction angles roughly bounded by

$$\beta \approx 1.6\theta + 4.7$$

and

$$\beta \approx 0.53\theta - 0.22$$

Discussion

It is important to consider the efficiency of a diffraction grating, as it can change by orders of magnitude within a few degrees of diffractive angle. The actual efficiency is a complicated function of polarization, blaze angle, reflection (or transmission), and diffraction angle.

The above is based on a curve fit for gratings of blaze angles from 2 to 26.5° and assumes the "P" polarization, as the "S" ray's efficiency can be a very complicated function of the diffraction angle and is generally much lower. This also assumes a first-order Littrow configuration and triangular grove geometry.

Figure 14.1 shows the data for an efficiency of 50 percent. For diffracted angles between the two lines, the efficiency will be greater than 50 percent.

For example, if the Blaze angle is 15°, the efficiency is above 50 percent from approximately

$$0.56(15) + 0.3 = 8.7°$$

to

$$1.1(15) + 2 = 18.5°$$

Thus, one can assume that they will have greater than 50 percent efficiency from about 9 to 19° of diffraction angle. Note that this rule, like the one that follows, is a crude approximation.

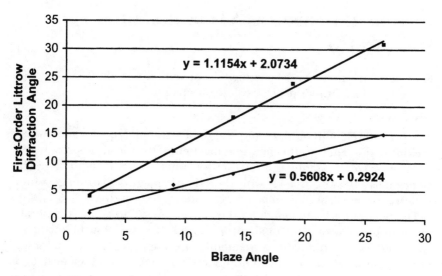

Figure 14.1 Fifty percent diffraction grating efficiency.

Reference

1. C. Palmer, *Diffraction Grating Handbook*, Thermo RGL, Rochester, NY, pp. 88–92, 2000.

GRATING EFFICIENCY AS A FUNCTION OF WAVELENGTH

Fifty percent efficiency is obtained from approximately $0.7\lambda_b$ to $1.8\lambda_b$ where λ_b = the grating's peak or (blaze wavelength).

Discussion

The peak wavelength or blaze wavelength of a grating is the location where the diffraction energy efficiency is at maximum. It typically approaches 98 percent or so (the same as a mirror) for a narrow wavelength. The efficiency falls off more quickly with shorter wavelengths than longer wavelengths, so the λ_b should not be in the center of the bandpass.

The blaze angle is the angle of the cut on the grating, and it generally varies from a few degrees to about 50° or 60°. For very low blaze angles, (say, < 50), polarization effects frequently can be ignored. However, for any blaze angle above 5°, polarization effects can make the efficiency curve complicated and must be carefully considered. The wavelength efficiency is a slight function of the blaze angle. For instance, Palmer[1] points out that, for blaze angles of less than 5°, the 50 percent efficiency is from $0.67\lambda_b$ to $1.8\ \lambda_b$, while for blaze angles $22° < 0 < 38°$, the 50 percent point is from $0.7\lambda_b$ to $2\lambda_b$.

This rule assumes the Littrow configuration. Note that this rule is a crude approximation.

The reader may notice a difference between this rule and the one on the previous page. These rules present different methods of indicating where the grating efficiency is high. But both are crude approximations, and the results may differ substantially.

References

1. C. Palmer, *Diffraction Grating Handbook*, Thermo RGL, Rochester, NY, pp. 85–93, 2000.
2. E. Loewen et al., "Grating Efficiency Theory as Applied to Blazed and Holographic Gratings," *Applied Optics* 16, pp. 2711–2721, 1977.
3. Sample grating efficiency curves showing these results also appear on the web site of World Precision Instruments, http://www.wpiinc.com/WPI_Web/Spectroscopy/Gratings.html.

GRATING GROOVE SIZE

A groove facet should be flat, with smooth straight edges, and be generally free from irregularities on a scale of <1/10 of the wavelength of light being diffracted.

Discussion

For DWDM applications, this indicates that any diffraction grating used should be free of defects larger than about 130 to 150 nm. It is possible to understand the reason for this type of rule when one remembers that gratings make use of reflection or refraction to disperse the light into superposed waves of common wavelength. It is also necessary to remember that a grating (imagine a blazed grating) is simply a flat piece of glass with successive spaces displaced by a small amount. In effect, it mimics a flat piece of glass within each of the grating facets. Now consider what happens when light encounters such a surface.

This process is not unlike what occurs when light is refracted by a lens or reflected by a mirror. That is, for the light to form a wavefront that is flat enough to create images or to be otherwise manipulated, the optical elements must not add distorting factors. Furthermore, it is typical that a lens or mirror system should have surface quality of about 1/10 wave to assure that a clean wavefront results. So, too, is the case with gratings.

Reference

1. C. Palmer, *Diffraction Grating Handbook*, Thermo RGL, Rochester, NY, p. 87, 2000.

KARTALOPOULOS' CHARACTERISTICS OF ARRAYED-WAVEGUIDE GRATINGS

Kartalopoulos[1] gives six salient characteristics of arrayed-waveguide gratings (AWGs) as paraphrased below.

1. AWGs are polarization dependent, but means to mitigate this dependency have been reported.

2. AWGs are temperature sensitive.

3. AWGs operating in wide temperature ranges have been reported (0 to 85°C).

4. AWGs exhibit good flat spectral response; this eases wavelength control.

5. Insertion loss is less than –3 dB, and crosstalk is better than –35 dB.

6. AWGs are suitable for integration with photodetectors.

Discussion

AWGs are typically made of SiO_2 or InP. These devices belong to the category of phased array gratings and waveguide grating routers. AWGs are based on interferometry in which light is dispersed into a cavity then routed through different length fibers (hence, different travel lengths and times) into another cavity, then multiplexed to an array of fibers. These are typically used as optical demultiplexers.

Their role in modern fiber systems is critical. They allow the many wavelengths needed for DWDM to be uniquely sensed in the detection process. Other rules in this chapter describe methods for getting the most out of these devices, both through their design and by managing their environment. The typical high-performance AWG is able to handle up to 40 simultaneous channels, each with a bandwidth of 100 GHz. The performance of AWG devices has also been developed to the point where isolation between channels is high (about 30 dB between any two channels), and noise performance is around –50 dB. The latter number is nearly at the limit of current measurement devices.

Reference

1. S. Kartalopoulos, *Introduction to DWDM Technology*, IEEE Press, Piscataway, NJ, p. 93–94, 2000.

MACH–ZEHNDER OPTICAL CHANNEL SPACING

The optical channel spacing from a Mach–Zehnder (MZ) interferometer (or filter) can be calculated from

$$\Delta f = \frac{c}{2n\Delta L}$$

where Δf = spacing between maximums in the frequency domain
  c = speed of light
  n = index of refraction
  ΔL = difference in travel paths in the same units as the velocity of light

Discussion

The equation will be recognized as one that appears in every discussion of Fabry–Perot etalons.

This is a fine place to start by showing how frequency and wavelength are related. We start by noting that $v = c/\lambda$. This provides the conversion to $\Delta\lambda$ $= |\ c/v^2\Delta v\ |$. When we put in some specific numbers, we can see how truly narrow 25 GHz is. At 1.55 microns, the frequency is 1.93×10^{14} Hz. A little math shows that 25 GHz converts to a wavelength spectral band of 0.2 nm, which is quite narrow indeed. Moreover, some companies are projecting going to 10 or 5 GHz spacing for streaming video. The complete grid for this and 50- and 100-GHz spacing can be found in Appendix A.

The optical channel spacing of a Mach–Zehnder filter is described by the equation. An MZ filter is an interferometer, which relies on a path-length difference to produce a series of wavelengths with the above spacing per the figure. A single input wavelength can be split, part of the amplitude (or brightness) sent down one fiber path length (L_1), and the other down a slightly different path length ($L_1 + \Delta L$). They are recombined, and their phase difference results in interference and an interferometric series of wavelengths with the above spacing. Additional filtering can filter these wavelengths to one or multiple wavelengths.

When recombined, the light from the alternative paths has a phase difference equal to

$$\Delta = \frac{2\pi f(\Delta L)n}{c}$$

where Δ = difference in phase between the two light paths of the above figure
  f = input frequency
  ΔL = path length difference

n = index of refraction

c = speed of light

The above equation results in a wavelength dependent pattern that peaks every $2m\pi$, where m is any positive integer. Thus, a plethora of wavelengths with the Δf spacing are produced by an MZ device, although the intensity usually limits the useful wavelengths to a few (small integer). However, a bandpass filter is needed to block any undesired wavelengths.

One advantage of the MZ scheme is that it can be easily tunable. An MZ interferometer (or filter) can be made tunable by simply varying the lengths that the light has to travel. If the path lengths are determined by the fiber, this can be accomplished by several means, including temperature control or mechanical compression (by a piezo). Additionally, this length can be changed by inserting a liquid crystal material with an index of refraction that changes due to applied current, or any number of other means. Figure 14.2 is a cartoon meant to convey the concept, but the wavelengths are not actually spatially stacked as shown.

An MZ filter has a transmission profile that is the inverse of that of a Fabry–Perot filter. Thus, when an FP tends to reject, an MZ tends to transmit (see Fig. 14.2).

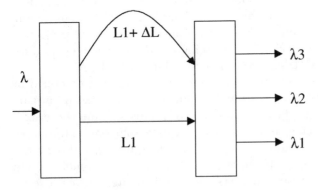

Figure 14.2

PULSE BROADENING IN A FABRY–PEROT ETALON

The final pulse width (in seconds) of a laser pulse after N trips through a Fabry Perot etalon is given by

$$\Delta t \approx 3.5 \times 10^{-11} F_d \sqrt{N}$$

where F = finesse of the etalon (which can be as large as ≈ 100)
d = thickness
N = number of trips

Discussion

Siegman[1] provides an exercise for those using his book (which includes a more complex representation of this formula) to come up with the simpler form presented above. The result applies to the pulse width after many trips through the etalon. Curiously, the more complex form includes the length of the initial pulse. This version does not.

This rule provides a simple method for estimating the length of a pulse that has passed through an etalon many times. Note that the pulse continues to grow without bound, since N is present under a square root.

Etalons provide extremely high-performance optical filters that include lasers, spectral filters, and science instruments. Their nice features include the fact that they have extremely narrow bandpass (within a range of wavelengths called the *free spectral range*), and they can be made quite robust mechanically. In fact, the equations that govern the performance of etalons are the same as those that define the performance of multilayer coatings used to create anti-reflection coatings and related optical devices. In typical applications, the etalon is able to resolve wavelengths down to hundredths of nanometers or smaller. In addition, they form the basis of most laser resonators and can be used to control the number of modes that propagate freely in the resonator. Most optics books provide a detailed discussion of the equations that define how etalons work and how to compute their resolving power.

Reference

1. A. Siegman, *Lasers*, University Science Books, Mill Valley, CA, p. 360, 1986.

RESOLVING POWER OF A GRATING

The resolving power of a grating is approximately 1000 times its length in millimeters.

Discussion

The resolving power of a grating is the reciprocal of the grating's diffraction-limited bandwidth. The resolving power is determined by the focusing optics and the flatness and quality of the grating, and there is much argument over its definition.

The above relationship can be 10 or 20 percent pessimistic for a well designed system. Additionally, the derivation of the above assumes that almost all of the lines of the grating are illuminated, and the wavelength is around 1500 to 2000 nm. The resolving power increases toward smaller wavelengths, so the grating will perform better in the C band than the L band, and even better in the older, shorter bands. Surface quality, however, is more demanding at the popular 1310-nm window.

Hobbs[1] gives a maximum resolving power of a grating of approximately two times the length divided by the wavelength. His equation would yield the above results for a wavelength of 2 microns.

Surface quality is an issue as well. The criteria for optical surface quality of a mirror and a grating is the scratch-and-dig figure and surface uniformity (wave aspect ratio). Finally, since gratings make use of reflection to disperse the light, polarization cannot be avoided. Systems sensitive to polarization dispersion must take this into account.

Reference

1. P. Hobbs, *Building Electro-Optical Systems, Making It All Work*, John Wiley & Sons, New York, p. 219, 2000.

STABILIZING WAVELENGTH OF ARRAYED-WAVEGUIDE GRATING

When bending an arrayed-waveguide grating (AWG) chip, the pitch of the grating is altered. This can be useful in stabilizing the wavelength of operation, but only within limits. The following equation quantifies the maximum stress (σ_{max}) that can be applied as a function of the chip's properties:

$$\sigma_{max} = \frac{1}{6t_f(t_f + t_s)}\frac{1}{R}\left\{ \frac{E_f t_f^3}{1 - v_f^2} + \frac{E_s t_s^3}{1 - v_s^2} \right\}$$

where t = thickness
E, v = Young's modulus and Poisson's ratio of materials, respectively (designated by f for the film and s for the substrate of the device)
R = radius of the substrate of the AWG

Discussion

The continued effort to create more bandwidth without installing more cables imposes demands on the spectral bandwidth of the carrier signals. The most common approach is to create DWDM systems operating with many wavelengths. A critical component for such operation is the multiplexer/demux component. An example of a critical component is the AWG module. This device exhibits compactness, support for many very small channels, and good producibility. A nagging problem, however, is that AWG performance demands careful temperature control to assure that the desired wavelengths are maintained. As the reference illustrates, it is possible to control the effects of temperature variation, assuring that the exquisite wavelength control requirements are maintained.

The curvature can be achieved by using a bimetal strip that is controlled by a temperature sensor, or by using any other mechanism for curving the substrate. Another approach using bimetallic stress is shown in Ref. 1.

Saito and Ota[2] claim to have solved this problem by including a stress management component that allows a control loop to stabilize the operating wavelength of the component. They can control the operating wavelength to within 0.03 nm over a temperature range of 70°C.

References

1. S. Lin, "Temperature Compensating Device with Central-Wavelength Tuning for Optical Fiber Gratings," *Optical Engineering* 40(5), p. 700, May 2001.
2. T. Saito and T. Ota, "Temperature Insensitive AWG Module by Stress Control," *Proceedings of the National Fiber Optic Engineers Conference*, 2000.

STRAIN IN FIBER BRAGG GRATINGS

Strain in fiber Bragg gratings (FBGs) can be used for spectral tuning or selection.

Discussion

Strain imposed along the length of a fiber Bragg grating can change both the pitch of the imposed variations in its index of refraction (which are the source of the grating feature itself) and the index of refraction of the material of the grating. This has the effect of selecting and deselecting wavelengths. This phenomena has been employed in some network elements as well as test tools.

Use the following equation to compute the variation of the Bragg wavelength (λ) with strain, ε. Λ is the pitch of the unstrained grating.

$$\frac{1}{\lambda}\frac{\Delta\lambda}{\varepsilon} = \frac{1}{\Lambda}\frac{\Delta\Lambda}{\varepsilon} + \frac{1}{n_{\text{eff}}}\frac{\Delta n_{\text{eff}}}{\varepsilon}$$

The equation shows that the change in wavelength ($\delta\Lambda$) depends on the properties of the grating, along with the changes in them that are induced by strain. The first term on the right side of the equation shows the impact of strain on the physical pitch of the grating, while the second term shows that the index of refraction of the material also changes under the influence of the strain.

Reference

1. R. Claus, I. Matias, and F. Arregui, "Optical Fiber Sensors," Chap. 15 in *Fiber Optics Handbook*, M. Bass, Ed. in Chief, and E. Van Stryland, Assoc. Ed., McGraw-Hill, New York, p. 15.8, 2001.

Appendix A
Useful Values and Conversions

The following collection of values represent frequently needed information and conversion factors for the EO telecom practitioner. This eclectic collection consist of constants, conversions, and definitions categorized by the authors' whims. Although sometimes supplied, the authors are not encouraging the use of discouraged and archaic units. Such units are included here only as a help with their translation to more accepted SI units.

Angular Measurement

Degree (angle)	0.01745 radians
	17.5 milliradians
	60 minutes
	3600 seconds
	17,452 microradians
Radian	0.159 of the circumference
	57.296 degrees
	3438 minutes
	2.06×10^5 seconds
Steradian	0.08 of total solid angle
RPM	6 degrees/second
	0.0167 revolutions/second
	0.105 radians/second
RPS	21,600 degrees/minute

Area Measurements

Square centimeter	1.076×10^{-3} feet^2
	0.155 inches^2
	1×10^{-4} meters^2
Square mil	645 microns^2
Square inch	6.452 centimeters^2
Square foot	939 centimeters^2
	0.092 meters^2

dBm to Watts

dBm to watts	1 pW = −90 dBm
	1 nW = −60 dBm
	1 µW = −30 dBm
	1 mW = −10 dBm
	10 mW = +10 dBm

Density Measurements

Density of water (4°C, 760 mm Hg)	1 g/centimeter^3
	1000 kilograms/meter^3
	62.43 pounds/foot^3
	0.036 pounds/inch^3
Grams/centimeter^3	0.036 pounds/inch^3
	64.2 pounds/foot^3
Pounds/inch^3	27.7 grams/centimeter^3

Electromagnetic Spectrum in Nanometers

Gamma rays	< 1
X-rays	1–20
Ultraviolet	≈20 to ≈400
Visible	≈400 to ≈750
Near infrared	≈750 to ≈1200
Short-wave infrared	≈1200 to 3000
Mid-wave infrared	≈3000 to 6000
Long-wave infrared	≈6000 to 14000
Far infrared	≈14,000 to 100,000
Submillimeter	≈100,000 to 1,000,000
Radio frequency	>1,000,000

Energy

Number of photons in a watt	$5.03 \times 10^{18} \times \lambda$ (with λ in microns)
BTU	252 calories 1055 joules 0.29 watt-hours
Calories	4.184 joules
Energy of one electron volt	1.602×10^{-19} J
Energy per photon	$1.98 \times 10^{-19}/\lambda$ watts seconds, where λ is in microns
Erg	1×10^{-7} joules 2.78×10^{-11} watt hours
Joule	9.48×10^{-4} BTU 0.2388 cal 1×10^{7} erg 1×10^{7} dyne centimeter 1 watt second 1 volt coulomb 0.738 foot pound 2.78×10^{-4} watt hours 1 newton meter 3.73×10^{-7} horsepower hours
Megajoule	1×10^{6} joules 2.4×10^{7} calories The kinetic energy of a Cadillac traveling 55 mph
Kilogram	9×10^{16} joules

Fiber Telecom Spectral Bands

Band nomenclature	Name	Wavelength (nm)
Fiber channel	Fiber channel	770 to 860
IEEE Serial Bus	IEEE Serial Bus	830 to 860
HIPPI	HIPPI	830 to 860 and 1260 to 1360
O	Original	1260 to 1360
S	Short	1360 to 1460
E	Extended	1460 to 1530
C	Conventional	1530 to 1565
L	Long	1565 to 1625
U	Ultra-long	1625 to 1675

Laser Lines (Popular) in Nanometers

Alexandrite	$\approx$720 to 800
Argon	$\approx$510
CO	$\approx$5000 to 7000
CO_2	$\approx$9200 to 11,000 $\approx$10,600
DF	$\approx$3800 to 4000
Doubled Nd:Yag	$\approx$530
Dy:CaF	$\approx$2350
Er:Yag	$\approx$1640
Erbium	1554 to 1556 (the popular eyesafe wavelength, often used)
GaAs	$\approx$800 to 1700
HeNe	$\approx$632.8 $\approx$594.4 to 614.3 $\approx$1152 $\approx$3391
HF	$\approx$2600 to 3000
Kr	$\approx$350
Nd: Yag	$\approx$1064.5
Ne laser	$\approx$332.4 $\approx$540.1
Nitrogen laser	$\approx$330
Ruby	$\approx$690

Length

Centimeter	0.3937 inch 1×10^4 micron 394 mils
Kilometer	3281 feet 0.54 nautical miles 0.621 statute miles 1094 yards

Meter	1×10^{10} angstroms
	3.28 feet
	39.37 inch
	1×10^9 nanometers
	1.094 yards
Mil	0.001 inch
	25.4 microns
	0.0254 millimeters
Inches	2.54 centimeters
Mile (nautical)	1852 meters
	1.15 statute miles
Mile (statute)	5280 feet
	63360 inches
	160,934 centimeters
	1.609 kilometers
	1609 meters

Miscellaneous

Weight of air	1.2 kilograms/meter^3
False alarm probability (in white noise for a point source) at a given SNR with a P_d of 0.99	SNR of 5, Pfa $\approx$ 0.01
	SNR of 6, Pfa $\approx 5 \times 10^{-4}$
	SNR of 7, Pfa $\approx 1 \times 10^{-5}$
	SNR of 9, Pfa $\approx 1 \times 10^{-10}$
Gaussian probability that a value will not exceed	1 sigma: 68.3%
	2 sigma 95.4%
	3 sigma 99.7%
Work month (average)	163 hours
Work week (average)	37.5 hours
Work year	$\approx$2000 hours
High region of the human eye's response	$\approx$0.4–0.65 microns
Water heat of fusion	80 gram-calories
Volume of one mole of gas at STP	22.4 liters

Numerical Constants

e	2.718281828459045
π	3.141592653589793

Optics

Amount of energy in circular diffraction pattern	84% in center disk an additional 7.1% in first bright ring an additional 2.8% in second bright ring an additional 1.5% in third bright ring
optical density $[= \log 10(T)]$	0 = 1.0 opacity and 100% transmission 0.5 = 3.2 opacity and 32% transmission 1.0 = 10.0 opacity and 10% transmission 1.5 = 32 opacity and 3.2% transmission 2.0 = 100 opacity and 1% transmission 3.0 = 1000 opacity and 0.1% transmission
Refractive index of glass	$\approx$1.5 to 2.0
Refractive index of quartz	$\approx$1.3 to 1.5
Refractive index of Si	$\approx$3.4

Photonic

Candle per foot^2	3.38×10^{-3} lamberts
Foot candle	1 lumen/foot^2 10.76 lumen/meter^2 10.76 lux
Lambert	0.318 lamberts/centimeter^2 295.7 candela/foot^2
Lumen per foot^2	10.76 lux
Lux	One lumen/meter^2 1×10^{-4} phot

Physical Constants

Atomic mass unit	1.657×10^{-24} grams
Avogadro's number	6.022×10^{-23} molecules/mole
Boltzmann's constant	1.3806×10^{-23} watt seconds/degree 1.3806×10^{-23} joules/kelvin
Charge of an electron	1.602×10^{-19} coulombs
Gravitational constant	6.67×10^{-11} m^3/kilogram seconds^2
Permeability of free space	1.2566×10^{-6} henries/meter

Permittivity of free space	8.854×10^{-12} farads/meter
Planck's constant	6.6252×10^{-34} watt seconds^2 or 6.6254×10^{-34} joule seconds
Mass of a neutron	1.675×10^{-27} kilograms
Mass of a proton	1.673×10^{-27} kilograms
Mass of an electron	9.109×10^{-31} kilograms
Velocity of light	2.99793×10^8 meters/second 2.99793×10^{10} centimeters/second 2.99793×10^{14} microns/second

Pressure

Dynes per centimeter^2	1.02×10^{-3} grams/centimeter^2 1.45×10^{-5} pounds/inch^2
One atmosphere	1.0133 bars 1.013×10^6 dynes/centimeter^2 760 Torr 1033 grams/centimeter^3 760 millimeters Hg
Ounce (mass)	28.35 grams
Pounds per inch^2	6895 pascals 0.068 atmospheres 51.71 Torr 51715 mm Hg
Torr	133.32 pascals 0.00133 bar 760 mm Hg

Receiver (Popular) Bandpasses in Nanometers

Germanium photodiode wavelength range	$\approx$800 to 1700
Indium gallium arsenide photodiode wavelength range	$\approx$900 to 1800
Silicon photodiode wavelength range	$\approx$500 to 1000

SONET/OC Rates

SONET	SDH	Optical	Data rate (Mb/s)	SONET	SDH	Optical	Data rate (Mb/s)
STS-1	STM-0	OC-1	51.84	STS-48	STM-16	OC-48	2488.32
STS-3	STM-1	OC-3	155.52	STS-192	STM-64	OC-192	9953.28
STS-12	STM-4	OC-12	622.08	STS-768	STM-254	OC-768	39813.12

Temperature

0°C	273 K 492° Rankine 32°F
0 K	−273.16°C −459.7°F
A difference of 1°C	1 K 1.8° Fahrenheit 1.8° Rankine 0.8° Re' aumur

Time

Hour	3600 seconds 0.04167 days 5.95×10^{-3} weeks
Day (mean = 24 hours)	1440 minutes 86400 seconds
Month (average)	30.44 days 730.5 hours 2.63×10^6 seconds 4.348 weeks
Year	365.256 days (sidereal) 8766 hours 525,960 minutes 3.16 million seconds 52.18 weeks

Weight

Gram	2.2×10^{-3} pounds 0.001 kilograms
Ounce (mass)	28.35 grams
Pound (mass)	454 grams

Example 100 GHz Grid for C and L Bands (Based on ITU Guidelines)

GHz	λ nm	Δλ	GHz	λ nm	Δλ	GHz	λ nm	Δλ
197100	1521.02		194400	1542.14	0.79	191700	1563.86	0.82
197000	1521.79	0.77	194300	1542.94	0.79	191600	1564.68	0.82
196900	1522.56	0.77	194200	1543.73	0.79	191500	1565.50	0.82
196800	1523.34	0.77	194100	1544.53	0.80	191400	1566.31	0.82
196700	1524.11	0.77	194000	1545.32	0.80	191300	1567.13	0.82
196600	1524.89	0.78	193900	1546.12	0.80	191200	1567.95	0.82
196500	1525.66	0.78	193800	1546.92	0.80	191100	1568.77	0.82
196400	1526.44	0.78	193700	1547.72	0.80	191000	1569.59	0.82
196300	1527.22	0.78	193600	1548.51	0.80	190900	1570.42	0.82
196200	1527.99	0.78	193500	1549.32	0.80	190800	1571.24	0.82
196100	1528.77	0.78	193400	1550.12	0.80	190700	1572.06	0.82
196000	1529.55	0.78	193300	1550.92	0.80	190600	1572.89	0.82
195900	1530.33	0.78	193200	1551.72	0.80	190500	1573.71	0.83
195800	1531.12	0.78	193100	1552.52	0.80	190400	1574.54	0.83
195700	1531.90	0.78	193000	1553.33	0.80	190300	1575.37	0.83
195600	1532.68	0.78	192900	1554.13	0.81	190200	1576.20	0.83
195500	1533.47	0.78	192800	1554.94	0.81	190100	1577.03	0.83
195400	1534.25	0.78	192700	1555.75	0.81	190000	1577.86	0.83
195300	1535.04	0.79	192600	1556.56	0.81	189900	1578.69	0.83
195200	1535.82	0.79	192500	1557.36	0.81	189800	1579.52	0.83
195100	1536.61	0.79	192400	1558.17	0.81	189700	1580.35	0.83
195000	1537.40	0.79	192300	1558.98	0.81	189600	1581.18	0.83
194900	1538.19	0.79	192200	1559.79	0.81	189500	1582.02	0.83
194800	1538.98	0.79	192100	1560.61	0.81	189400	1582.85	0.84
194700	1539.77	0.79	192000	1561.42	0.81	189300	1583.69	0.84
194600	1540.56	0.79	191900	1562.23	0.81	189200	1584.53	0.84
194500	1541.35	0.79	191800	1563.05	0.81	189100	1585.36	0.84

GHz	λ nm	Δλ	GHz	λ nm	Δλ	GHz	λ nm	Δλ
189000	1586.20	0.84	187600	1598.04	0.85	186200	1610.06	0.86
188900	1587.04	0.84	187500	1598.89	0.85	186100	1610.92	0.87
188800	1587.88	0.84	187400	1599.75	0.85	186000	1611.79	0.87
188700	1588.73	0.84	187300	1600.60	0.85	185900	1612.65	0.87
188600	1589.57	0.84	187200	1601.46	0.86	185800	1613.52	0.87
188500	1590.41	0.84	187100	1602.31	0.86	185700	1614.39	0.87
188400	1591.26	0.84	187000	1603.17	0.86	185600	1615.26	0.87
188300	1592.10	0.85	186900	1604.03	0.86	185500	1616.13	0.87
188200	1592.95	0.85	186800	1604.88	0.86	185400	1617.00	0.87
188100	1593.79	0.85	186700	1605.74	0.86	185300	1617.88	0.87
188000	1594.64	0.85	186600	1606.61	0.86	185200	1618.75	0.87
187900	1595.49	0.85	186500	1607.47	0.86	185100	1619.62	0.87
187800	1596.34	0.85	186400	1608.33	0.86	185000	1620.50	0.88
187700	1597.19	0.85	186300	1609.19	0.86			

Example 50 GHz Grid for C and L Bands (Based on ITU Guidelines)

GHz	λ nm	GHz	λ nm	GHz	λ nm
197100	1521.02	195750	1531.51	194400	1542.14
197050	1521.40	195700	1531.90	194350	1542.54
197000	1521.79	195650	1532.29	194300	1542.94
196950	1522.18	195600	1532.68	194250	1543.33
196900	1522.56	195550	1533.07	194200	1543.73
196850	1522.95	195500	1533.47	194150	1544.13
196800	1523.34	195450	1533.86	194100	1544.53
196750	1523.72	195400	1534.25	194050	1544.92
196700	1524.11	195350	1534.64	194000	1545.32
196650	1524.50	195300	1535.04	193950	1545.72
196600	1524.89	195250	1535.43	193900	1546.12
196550	1525.27	195200	1535.82	193850	1546.52
196500	1525.66	195150	1536.22	193800	1546.92
196450	1526.05	195100	1536.61	193750	1547.32
196400	1526.44	195050	1537.00	193700	1547.72
196350	1526.83	195000	1537.40	193650	1548.12
196300	1527.22	194950	1537.79	193600	1548.51
196250	1527.61	194900	1538.19	193550	1548.92
196200	1527.99	194850	1538.58	193500	1549.32
196150	1528.38	194800	1538.98	193450	1549.72
196100	1528.77	194750	1539.37	193400	1550.12
196050	1529.16	194700	1539.77	193350	1550.52
196000	1529.55	194650	1540.16	193300	1550.92
195950	1529.94	194600	1540.56	193250	1551.32
195900	1530.33	194550	1540.95	193200	1551.72
195850	1530.73	194500	1541.35	193150	1552.12
195800	1531.12	194450	1541.75	193100	1552.52

GHz	λ nm	GHz	λ nm	GHz	λ nm
193050	1552.93	191650	1564.27	190250	1575.78
193000	1553.33	191600	1564.68	190200	1576.20
192950	1553.73	191550	1565.09	190150	1576.61
192900	1554.13	191500	1565.50	190100	1577.03
192850	1554.54	191450	1565.90	190050	1577.44
192800	1554.94	191400	1566.31	190000	1577.86
192750	1555.34	191350	1566.72	189950	1578.27
192700	1555.75	191300	1567.13	189900	1578.69
192650	1556.15	191250	1567.54	189850	1579.10
192600	1556.56	191200	1567.95	189800	1579.52
192550	1556.96	191150	1568.36	189750	1579.93
192500	1557.36	191100	1568.77	189700	1580.35
192450	1557.77	191050	1569.18	189650	1580.77
192400	1558.17	191000	1569.59	189600	1581.18
192350	1558.58	190950	1570.01	189550	1581.60
192300	1558.98	190900	1570.42	189500	1582.02
192250	1559.39	190850	1570.83	189450	1582.44
192200	1559.79	190800	1571.24	189400	1582.85
192150	1560.20	190750	1571.65	189350	1583.27
192100	1560.61	190700	1572.06	189300	1583.69
192050	1561.01	190650	1572.48	189250	1584.11
192000	1561.42	190600	1572.89	189200	1584.53
191950	1561.83	190550	1573.30	189150	1584.95
191900	1562.23	190500	1573.71	189100	1585.36
191850	1562.64	190450	1574.13	189050	1585.78
191800	1563.05	190400	1574.54	189000	1586.20
191750	1563.46	190350	1574.95	188950	1586.62
191700	1563.86	190300	1575.37	188900	1587.04

GHz	λ nm	GHz	λ nm	GHz	λ nm
188850	1587.46	187700	1597.19	186550	1607.04
188800	1587.88	187650	1597.62	186500	1607.47
188750	1588.30	187600	1598.04	186450	1607.90
188700	1588.73	187550	1598.47	186400	1608.33
188650	1589.15	187500	1598.89	186350	1608.76
188600	1589.57	187450	1599.32	186300	1609.19
188550	1589.99	187400	1599.75	186250	1609.62
188500	1590.41	187350	1600.17	186200	1610.06
188450	1590.83	187300	1600.60	186150	1610.49
188400	1591.26	187250	1601.03	186100	1610.92
188350	1591.68	187200	1601.46	186050	1611.35
188300	1592.10	187150	1601.88	186000	1611.79
188250	1592.52	187100	1602.31	185950	1612.22
188200	1592.95	187050	1602.74	185900	1612.65
188150	1593.37	187000	1603.17	185850	1613.09
188100	1593.79	186950	1603.60	185800	1613.52
188050	1594.22	186900	1604.03	185750	1613.96
188000	1594.64	186850	1604.46	185700	1614.39
187950	1595.07	186800	1604.88	185650	1614.83
187900	1595.49	186750	1605.31	185600	1615.26
187850	1595.91	186700	1605.74	185550	1615.70
187800	1596.34	186650	1606.17	185500	1616.13
187750	1596.76	186600	1606.61	185450	1616.57

Example 25 GHz Grid for C and L Bands (Based on ITU Guidelines)

GHz	λ nm	GHz	λ nm	GHz	λ nm
197100	1521.02	196425	1526.24	195750	1531.51
197075	1521.21	196400	1526.44	195725	1531.70
197050	1521.40	196375	1526.63	195700	1531.90
197025	1521.60	196350	1526.83	195675	1532.09
197000	1521.79	196325	1527.02	195650	1532.29
196975	1521.98	196300	1527.22	195625	1532.49
196950	1522.18	196275	1527.41	195600	1532.68
196925	1522.37	196250	1527.61	195575	1532.88
196900	1522.56	196225	1527.80	195550	1533.07
196875	1522.76	196200	1527.99	195525	1533.27
196850	1522.95	196175	1528.19	195500	1533.47
196825	1523.14	196150	1528.38	195475	1533.66
196800	1523.34	196125	1528.58	195450	1533.86
196775	1523.53	196100	1528.77	195425	1534.05
196750	1523.72	196075	1528.97	195400	1534.25
196725	1523.92	196050	1529.16	195375	1534.45
196700	1524.11	196025	1529.36	195350	1534.64
196675	1524.30	196000	1529.55	195325	1534.84
196650	1524.50	195975	1529.75	195300	1535.04
196625	1524.69	195950	1529.94	195275	1535.23
196600	1524.89	195925	1530.14	195250	1535.43
196575	1525.08	195900	1530.33	195225	1535.63
196550	1525.27	195875	1530.53	195200	1535.82
196525	1525.47	195850	1530.73	195175	1536.02
196500	1525.66	195825	1530.92	195150	1536.22
196475	1525.86	195800	1531.12	195125	1536.41
196450	1526.05	195775	1531.31	195100	1536.61

GHz	λ nm	GHz	λ nm	GHz	λ nm
195075	1536.81	194375	1542.34	193675	1547.92
195050	1537.00	194350	1542.54	193650	1548.12
195025	1537.20	194325	1542.74	193625	1548.32
195000	1537.40	194300	1542.94	193600	1548.51
194975	1537.59	194275	1543.13	193575	1548.71
194950	1537.79	194250	1543.33	193550	1548.92
194925	1537.99	194225	1543.53	193525	1549.12
194900	1538.19	194200	1543.73	193500	1549.32
194875	1538.38	194175	1543.93	193475	1549.52
194850	1538.58	194150	1544.13	193450	1549.72
194825	1538.78	194125	1544.33	193425	1549.92
194800	1538.98	194100	1544.53	193400	1550.12
194775	1539.17	194075	1544.72	193375	1550.32
194750	1539.37	194050	1544.92	193350	1550.52
194725	1539.57	194025	1545.12	193325	1550.72
194700	1539.77	194000	1545.32	193300	1550.92
194675	1539.96	193975	1545.52	193275	1551.12
194650	1540.16	193950	1545.72	193250	1551.32
194625	1540.36	193925	1545.92	193225	1551.52
194600	1540.56	193900	1546.12	193200	1551.72
194575	1540.76	193875	1546.32	193175	1551.92
194550	1540.95	193850	1546.52	193150	1552.12
194525	1541.15	193825	1546.72	193125	1552.32
194500	1541.35	193800	1546.92	193100	1552.52
194475	1541.55	193775	1547.12	193075	1552.73
194450	1541.75	193750	1547.32	193050	1552.93
194425	1541.94	193725	1547.52	193025	1553.13
194400	1542.14	193700	1547.72	193000	1553.33

GHz	λ nm	GHz	λ nm	GHz	λ nm
192975	1553.53	192275	1559.19	191575	1564.88
192950	1553.73	192250	1559.39	191550	1565.09
192925	1553.93	192225	1559.59	191525	1565.29
192900	1554.13	192200	1559.79	191500	1565.50
192875	1554.34	192175	1560.00	191475	1565.70
192850	1554.54	192150	1560.20	191450	1565.90
192825	1554.74	192125	1560.40	191425	1566.11
192800	1554.94	192100	1560.61	191400	1566.31
192775	1555.14	192075	1560.81	191375	1566.52
192750	1555.34	192050	1561.01	191350	1566.72
192725	1555.55	192025	1561.22	191325	1566.93
192700	1555.75	192000	1561.42	191300	1567.13
192675	1555.95	191975	1561.62	191275	1567.34
192650	1556.15	191950	1561.83	191250	1567.54
192625	1556.35	191925	1562.03	191225	1567.75
192600	1556.56	191900	1562.23	191200	1567.95
192575	1556.76	191875	1562.44	191175	1568.16
192550	1556.96	191850	1562.64	191150	1568.36
192525	1557.16	191825	1562.84	191125	1568.57
192500	1557.36	191800	1563.05	191100	1568.77
192475	1557.57	191775	1563.25	191075	1568.98
192450	1557.77	191750	1563.46	191050	1569.18
192425	1557.97	191725	1563.66	191025	1569.39
192400	1558.17	191700	1563.86	191000	1569.59
192375	1558.38	191675	1564.07	190975	1569.80
192350	1558.58	191650	1564.27	190950	1570.01
192325	1558.78	191625	1564.47	190925	1570.21
192300	1558.98	191600	1564.68	190900	1570.42

GHz	λ nm	GHz	λ nm	GHz	λ nm
190875	1570.62	190175	1576.40	189475	1582.23
190850	1570.83	190150	1576.61	189450	1582.44
190825	1571.03	190125	1576.82	189425	1582.64
190800	1571.24	190100	1577.03	189400	1582.85
190775	1571.45	190075	1577.23	189375	1583.06
190750	1571.65	190050	1577.44	189350	1583.27
190725	1571.86	190025	1577.65	189325	1583.48
190700	1572.06	190000	1577.86	189300	1583.69
190675	1572.27	189975	1578.06	189275	1583.90
190650	1572.48	189950	1578.27	189250	1584.11
190625	1572.68	189925	1578.48	189225	1584.32
190600	1572.89	189900	1578.69	189200	1584.53
190575	1573.09	189875	1578.89	189175	1584.74
190550	1573.30	189850	1579.10	189150	1584.95
190525	1573.51	189825	1579.31	189125	1585.16
190500	1573.71	189800	1579.52	189100	1585.36
190475	1573.92	189775	1579.73	189075	1585.57
190450	1574.13	189750	1579.93	189050	1585.78
190425	1574.33	189725	1580.14	189025	1585.99
190400	1574.54	189700	1580.35	189000	1586.20
190375	1574.75	189675	1580.56	188975	1586.41
190350	1574.95	189650	1580.77	188950	1586.62
190325	1575.16	189625	1580.98	188925	1586.83
190300	1575.37	189600	1581.18	188900	1587.04
190275	1575.57	189575	1581.39	188875	1587.25
190250	1575.78	189550	1581.60	188850	1587.46
190225	1575.99	189525	1581.81	188825	1587.67
190200	1576.20	189500	1582.02	188800	1587.88

GHz	λ nm	GHz	λ nm	GHz	λ nm
188775	1588.09	188075	1594.01	187375	1599.96
188750	1588.30	188050	1594.22	187350	1600.17
188725	1588.52	188025	1594.43	187325	1600.39
188700	1588.73	188000	1594.64	187300	1600.60
188675	1588.94	187975	1594.85	187275	1600.81
188650	1589.15	187950	1595.07	187250	1601.03
188625	1589.36	187925	1595.28	187225	1601.24
188600	1589.57	187900	1595.49	187200	1601.46
188575	1589.78	187875	1595.70	187175	1601.67
188550	1589.99	187850	1595.91	187150	1601.88
188525	1590.20	187825	1596.13	187125	1602.10
188500	1590.41	187800	1596.34	187100	1602.31
188475	1590.62	187775	1596.55	187075	1602.53
188450	1590.83	187750	1596.76	187050	1602.74
188425	1591.04	187725	1596.98	187025	1602.95
188400	1591.26	187700	1597.19	187000	1603.17
188375	1591.47	187675	1597.40	186975	1603.38
188350	1591.68	187650	1597.62	186950	1603.60
188325	1591.89	187625	1597.83	186925	1603.81
188300	1592.10	187600	1598.04	186900	1604.03
188275	1592.31	187575	1598.25	186875	1604.24
188250	1592.52	187550	1598.47	186850	1604.46
188225	1592.73	187525	1598.68	186825	1604.67
188200	1592.95	187500	1598.89	186800	1604.88
188175	1593.16	187475	1599.11	186775	1605.10
188150	1593.37	187450	1599.32	186750	1605.31
188125	1593.58	187425	1599.53	186725	1605.53
188100	1593.79	187400	1599.75	186700	1605.74

GHz	λ nm	GHz	λ nm	GHz	λ nm
186675	1605.96	186250	1609.62	185825	1613.31
186650	1606.17	186225	1609.84	185800	1613.52
186625	1606.39	186200	1610.06	185775	1613.74
186600	1606.61	186175	1610.27	185750	1613.96
186575	1606.82	186150	1610.49	185725	1614.17
186550	1607.04	186125	1610.71	185700	1614.39
186525	1607.25	186100	1610.92	185675	1614.61
186500	1607.47	186075	1611.14	185650	1614.83
186475	1607.68	186050	1611.35	185625	1615.04
186450	1607.90	186025	1611.57	185600	1615.26
186425	1608.11	186000	1611.79	185575	1615.48
186400	1608.33	185975	1612.00	185550	1615.70
186375	1608.54	185950	1612.22	185525	1615.91
186350	1608.76	185925	1612.44	185500	1616.13
186325	1608.98	185900	1612.65	185475	1616.35
186300	1609.19	185875	1612.87	185450	1616.57
186275	1609.41	185850	1613.09	185425	1616.79

Glossary

This glossary has been developed specifically to aid the reader of this volume in interpreting unfamiliar terms. The definitions are intended to have a practical, rather than formal, presentation. That is, they may be specifically oriented toward the use of the word in the telecommunications or electro-optics sciences rather than a more general definition that would be found in a generic encyclopedia or dictionary.

μrad	Microradian, a millionth of a radian.
μW	Microwatt, a millionths of a watt.
1000BaseT	A terabit per second over a twisted pair, per Ethernet Standard IEEE802.3ab.
aberrated	Defines specific degradations in image quality. Aberrations include, but are not limited to, field curvature, distortion, coma, chromatic aberration, and astigmatism.
absorptance	The property of a material to that acts to reduce the amount of radiation traversing through a section of the material. It is measured as the fraction of radiation absorbed traversing the material. Generally, the bulk absorptance of radiation by a material follows Beer's law.
acceptance angle	The half angle of the acceptance cone, this is the angle (or less) in which light will be easily coupled into a fiber.
adaptive optics	Optical subsystems with the capability to change the wavefront in real time.
ADM	Add-drop multiplexer.
ADSL	Asymmetric digital subscriber line.

afocal
Describes an optical system that does not form a focus, such as often employed in astronomical telescopes and in laser beam expanders. Such optics accept a light field and produce an unfocused beam.

Airy disk
The distribution of radiation at the ideal focus of a circular aperture in an optical system when dominated by diffraction effects (e.g., no aberrations). The central spot formed by about 84 percent of the total energy.

Al
Aluminum.

AN
Access node.

Angstrom (Å)
A unit of length, equal to 10^{-10} m, used to define wavelength, although it has been generally replaced by microns (10^4 Å) or nanometers (10^1 Å).

ANSI
American National Standards Institute.

antireflection coating
A coating intended to reduce the amount of radiation reflected from surface of an optical element.

AOTF
Acoustic optical tunable filter.

APD
Avalanche photodiode.

apodization
The modification of the transmission properties of an aperture to suppress unwanted optical aberrations or diffraction effects.

APS
Automatic protection switching.

areance
See *radiant sterance* and *luminous sterance*.

ASE
Amplified spontaneous emission.

ASIC
Application specific integrated circuit. These are custom-made chips "hardwired" to perform specific functions. They provide the lightest weight and lowest power processing but usually have little or no reprogramming ability.

aspheric
Not having a spherical shape. These shapes are often used in optics to define lenses and mirrors that are conic sections.

ATM
Asynchronous transfer mode. A switching and multiplexing mode using cells that are 53 octets in length.

avalanche photodiode
A semiconductor detector device that exploits the phenomena of photoemission and self-amplification.

B8zs
Bipolar eight-zero substitution.

background   The part of a sensed signal or scene that is not the item of interest (e.g., the ocean is the background for a system trying to detect swimming shipwreck survivors).

backscatter   That part of light impinging on a surface that is not absorbed, transmitted, or specularly reflected. It may be used to describe the part of the light scattered from a surface exposed to laser light, with the proviso that the backscattered light is that which is scattered toward the illuminating source.

baffle   A structural shield that prevents unwanted light from impinging on optics or the focal plane. Cold shields and sunshades are forms of baffles.

band gap   The energy band gap (often specified in electron-volts, eV) in a semiconductor material is the amount of energy needed to sufficiently interact with the lattice to generate free carriers, which are the source of the detected electrical signal. For photovoltaic and photoconductive devices, the band gap (once converted from eV to wavelength) defines the longest wavelength to which the detector is sensitive.

bandpass   The range of wavelengths over which a system functions. The term is also used in an electronic sense to define the range of electrical frequencies that are accommodated by an electrical system.

bend radius   The radius of a curve in which a fiber can be bent before signal loss or fiber breakage occurs.

BCH   Bose-Chaudhuri-Hocquenghem, a widely used technique to provide forward error correction.

BER   Bit error rate, the number of bits that are incorrect as compared with the number transmitted. A bit error rate of 1×10^{-6} means that one in a million are incorrect. The BER often is assumed to be on the order of 1×10^{-12}.

Bessel   A prominent mathematician whose work has been recognized by assigning his name to a function frequently used in defining the electromagnetic properties of circular apertures, such as in diffraction theory.

BICI   Broadband intercarrier interface.

bidirectional   Capable of transmitting in both directions.

binary optics   Optics that utilize diffraction to alter light rays. They are made from a photolithography (mask-and-etch) process. The photolithographic process results in the optical curve being approximated in a series of steps. This staircase approximation to a curve has the number of steps being a power of 2 (e.g., 2 steps, 4 steps, 16 steps, etc.); hence the use of the term *binary.*

BIP   Bit interleaved parity.

BLSR   Bidirectional line switched ring.

BPF   Bandpass filter.

C   Degrees C or Celsius, a scale developed by Swedish astronomer Andres Celsius in 1742, which has 100 divisions or degrees between the freezing point of water and the boiling point.

CAP   Competitive access provider.

Cassegrain   Any of a class of two-mirror telescopes in which the primary mirror is concave of a parabolic shape; the secondary mirror is convex and hyperbolic. These systems exhibit no spherical aberrations.

CATV   Cable television.

CBR   Constant bit rate.

CCD   Charge coupled device. A particular implementation of detector and/or readout technology composed of an array of capacitors. Charge is accumulated on a capacitor and shift-registered along rows and columns to provide a readout. These are frequently used in commercial video cameras, visible arrays, and as a readout multiplexer for infrared focal planes.

CCITT   Comité Consultatif International Téléphonique et Télégraphique (Consultative Committee of International Telegraph and Telephone). Now known by the name of the parent organization, International Telecommunication Union (ITU).

CD   Chromatic dispersion or compact disc.

CDC   Chromatic dispersion coefficient.

centroid   The resultant "center of mass" or other characterization of a distribution of light falling on an array of detectors. In the context of this book, it usually refers to light on a focal plane, often used to compute the location of a target to higher resolution than the angular size of a single pixel in the array.

chirp A characteristic of the laser diode in which the center frequency shifts during the pulse—generally an undesirable property for optical telecom.

CIE Commission Internationale de l'Eclairage, or the International Commission on Illumination, a standards-generating group concerned with color and illumination.

CIU Channel interface unit.

cladding The lower-index-material sheathing covering a fiber.

CLR Cell loss rate.

CMOS Complementary metal oxide semiconductor, a common architecture for integrated circuits.

CNR Carrier-to-noise ratio.

confocal The property of two or more optical components having the same focal point. Commonly used in describing certain types of laser resonators.

core The "glass" transmitting part of a fiber of a higher refractive index than the cladding. Commonly, the core is fused silica.

coupling loss The optical power loss of a junction (usually expressed in decibels of watts).

CPDWM Chirped-pulse wavelength division multiplexed.

CPE Customer premise equipment.

CRT Cathode ray tube, an old type of display.

CSF Cutoff shifted fiber (ITU-T: G.654); has its cutoff wavelength shifted from 1260 nm to over 1310 nm to ease transmissions in the L band.

CTE Coefficient of thermal expansion.

CU Channel unit.

CV Coding violation.

dark fiber Often meant to indicate a fiber that is not being used, sometimes in reference to a type of fiber installation in which the user must terminate the fiber.

dB Decibel.

dBm Decibel in reference to a 1-mW laser.

DBR	Distributed Bragg reflector.
DCF	Dispersion-compensating fiber.
decibel	An expression of a ratio equal to ten times the log of the ratio. Generally used for power gains or losses and electrical signal-to-noise ratios.
DFB	Distributed feedback diode laser.
DFF	Dispersion-flattened fiber.
DGD	Differential group delay.
die	An integrated circuit as discriminated on a wafer; the chip as delineated on the wafer. Plural is *dice*. After the circuit is "sawed" or cut from the wafer, it is called a "chip."
DPA	Dynamic packet assignment.
DPSS	Diode-pumped solid state laser.
DSF	dispersion-shifted fiber.
DSL	Digital subscriber line. Suffers from many reliability and customer care issues.
DWDM	Dense wavelength division multiplexing. Numerous lasers of slightly different wavelength (or frequencies) are launched into the same fiber. Generally meant to indicate more than 20 lasers, with a separation of 5 to 100 GHz.
DWE	Dynamic wavelength equalizer.
EBC	Error blocked count.
EC	Echo cancellation.
EDFA	Erbium-doped fiber amplifier.
EFL	Effective focal length. The resultant focal length of a system of lenses or mirrors that are working in conjunction.
EIA	Electronic Industries Alliance (formerly Electronic Industries Association).
ensquared	Usually used to describe the amount of energy focused into a square area (such as a detector). A similar term, *encircled*, defines the same type of measure for circular spots.
EO	Electro-optics. The field of study that integrates electronics and optics to create hybrid systems.
EOTF	Electro-optical tunable filter.

erf	The error function, defined in any book on mathematical physics or engineering physics.
ESA	Electrical spectrum analyzer.
etalon	A type of interferometer in which high spectral resolution can be obtained. The Fabry–Perot interferometer is such a device.
etendue	The product of an optical system's solid angle field of view and its effective collective area.
exitance	Flux density emitted from a surface.
eye pattern	A simultaneously overlayed display of the rise and fall of a "one" and "zero." This generally results in a pattern in the shape of an eye (oval) that indicates the bit error rate.
F	Fluoride.
F/#	Pronounced "f number," it is commonly defined the ratio of the effective focal length to the effective aperture.
Fabry	An optical scientist of about 100 years ago whose name is most often seen with that of Perot, as in the Fabry–Perot interferometer.
false alarm rate	The frequency at which a system generates false alarms. Related to the probability of detection and the amount of noise and clutter.
FBG	Fiber Bragg grating.
FDDI	Fiber distributed data interface.
FEC	Forward error correction.
ferrule	The protective tube that surrounds a fiber.
figure	The general overall shape of an optic (e.g., a parabola, sphere, etc.).
Fizeau	A prominent, early optical scientist whose name is associated with certain types of interferometers.
focal length	The distance from the principal plane to the focal plane.
Foucault	An important scientist who contributed to the study of both mechanics and optics, in the latter case especially related to the problem of determining the speed of light. The knife-edge test is named after him.
Fourier	A prominent scientist and mathematician who gives his name to any number of important fields of current study. In electro-optics, his name appears associ-

	ated with the field of Fourier optics, the Fourier transform, etc.
FOV	Field of view.
FRP	Fiber reinforced plastic.
FSO	Free-space optics (not using fibers for data transmission over a range).
FWHM	Full width, half maximum. A common measure of the width of a distribution of some function. For example, in laser physics, this term is often used to define the lateral dimension of a beam by determining the diameter that best meets the criterion of being at the half power point of the beam.
FWM	Four-wave mixing.
GaAs	Gallium arsenide. A semiconductor material used in electronics, lasers, optics, and detectors.
GEF	Gain equalization filter.
GHz	Gigahertz = 10^9 Hertz.
gradient index fiber	A fiber whose core index decreases in increasing radial distance from its center.
GRIN	Gradient index of refraction.
HeNe	Short for a helium neon, a type of laser. HeNe lasers are common, low in cost, and commercially available. They typically have a frequency of 0.6328 µm.
Hz	Hertz, cycles per second.
IEEE	Institute of Electrical and Electronics Engineers.
illuminance	Flux that is incident on a surface.
in-band	Having the property of being detectable by a particular electro-optic system. That is, the radiation is said to be within the wavelength limits of the system.
index of refraction	A property of all transparent or semi-transparent materials, defined by the speed of light in the material and, in turn, defining the effect of the material on the deviation of light entering from another medium of different index.
infrared	The portion of the electromagnetic radiation spectrum that extends in wavelength from beyond the visible ($\approx$0.8 µm) to the submillimeter regime ($\approx$100 µm).
InGaAs	Indium gallium arsenide; an important detector material in modern telecom, as it has a very high response in the C and L bands.

InP Indium phosphide; a common III-V compound used in optical telecom components.

interferometry The field of study of the development and application of systems that work by combining separate beams of light.

IP Internet protocol, indium phosphide, or intellectual property.

ISO International Standards Institute. This is the organization responsible for the various versions of the ISO quality standards, such as ISO9001 and ISO9002.

isoplanatic angle The range of angles over which the wavefront of a system is essentially constant.

K The abbreviation for kelvins, the accepted scientific unit of temperature.

kelvins The SI and internationally accepted metric unit of temperature and the most common measure of temperature in the electro-optics industry. Zero Kelvins is absolute zero. Each kelvin increment is the same as a degree centigrade. Note that the internationally accepted use of the term eliminates the word *degrees*, such as "the instrument operates at 10 kelvin (or kelvins)" and not "10 degrees kelvin."

LADAR Laser radar. A system for determining the range and sometimes the shape of distant targets by illuminating them with laser and light and detecting the reflected light with a co-located detection system. Bistatic systems exist wherein the receiver and transmitter are not co-located. Used interchangeably with LIDAR by most researchers, the latter meaning *light detection and ranging*.

lambert A unit of light fluence in the photopic/scotopic system of measurement. A lambert is defined as 3183.1 candles/m^2.

LAN Local area network.

launch angle The angle between the axis of a fiber and the propagating light.

L-band The DWDM band from 1565 to 1625 nm.

least significant bit (LSB) The smallest amount of information that is represented when an analog system is digitized. At best, it is about 1/2 of a bit from the A/D. This can be the dominant noise source for quiet, large-dynamic-range sensors.

LED	Light emitting diode; an important technology for producing the light used in old communications systems and displays.
lenslet	Usually used to define the properties of the lens systems employed in Shack–Hartmann wavefront sensors.
lightweighting	The act of removing unnecessary mass from optical systems.
LiNbO$_3$	Lithium niobate.
line of sight	The projection of the optic axis into object space. This represents the center of the field of view and where the sensor is pointed.
LiTaO$_3$	A compound (of lithium, tantalum, and oxygen) useful for low-sensitivity detection of infrared radiation employing the pyroelectric effect.
LSB	Least significant bit, see above.
luminous areance	lumen/meter^2.
luminous pointance	lumen/steradians.
luminous sterance	lumen/meter^2/steradian.
lux	A measure of radiation in the photopic/scotopic radiation measurement system; abbreviated lx.
LWIR	Long-wave infrared; the portion of the electromagnetic spectrum that is longer in wavelength than mid infrared and shorter in wavelength than very long infrared. It is typically defined by available detector technology to be from about 6 µm (where Pt:Si and InSb don't work) to $\approx$ 26 µm.
Mach–Zehnder	A type of interferometer characterized by two couplers connected by two waveguides of different optical lengths. Frequently used as a filter in DWDM applications.
magnetostriction	A phenomenon wherein a material changes dimensions as a function of magnetic fields (in applications, similar to electrostrictive and piezoelectric effects).
MAN	Metropolitan area network.
manufacturability	The measure of the ease with which something can be manufactured. It is usually included in the description of the risk associated with a particular EO system

or technology, since it implies a cost in either time or money.

meridional A plane through an optical system that includes the optic axis.

microlens A tiny lens, usually mounted directly to a focal plane.

micron A millionth of a meter. Optical communications occur between 0.8 and 1.7 microns.

microradian A millionth of a radian. Commonly abbreviated as μradian or μrad.

milliradian A thousandth of a radian. Commonly abbreviated as mrad.

milliseconds A thousandth of a second.

milliwatt A thousandth of a watt. Commonly abbreviated as mW.

MLM Multilongitudinal mode.

MMF Multimode fiber.

monochromatic Consisting of one color, or very narrow in wavelength bandpass.

MOPA Master oscillator power assembly when in reference to a laser.

MQW Multiple quantum well. Usually in reference to a MQW laser. Quantum well receivers or detectors are usually referred to as QWIPs.

MTF Modulation transfer function; a way to describe image quality.

MTTF Mean time to failure. A statistical measure indicating the reliability of a system, based on the time at which half of the units have failed.

multimode Generally, this refers to fiber optics or lasers and indicates that they are capable of transmitting (in the case of fibers) or generating (for lasers) radiation of more than one electromagnetic mode.

multiplexer An electronic system able to provide inputs from several sources to one operational module, each in turn. Often used in describing the design of electronic focal plane readout systems.

mux Multiplexer.

MWIR	Mid-wave infrared; usually includes radiation from about 3 to 6 μm.
MZ	A Mach–Zehnder interferometer, see Mach–Zehnder.
NA	Numerical aperture.
nanoamp	One billionth of an ampere.
NBA	Nano-Business Alliance.
NDSF	Nondispersion-shifted fiber; a fiber meant to operate in a low dispersion window near 1300 nm.
NE	Network element.
NEP	Noise equivalent power. A measure of the performance of a detector or system in which all of the noise terms are aggregated into one measure that produces a system signal-to-noise ratio of 1. NEP is defined with units of watts.
Newton	Unit of force in the metric system.
NFEOC	National Fiber Optic Engineers Conference, a major conference in optical telecom, held each summer.
nm	Nanometer. A commonly used measure of wavelength, defined as one billionth of a meter.
nonisotropic	Not having the properties of isotropy; that is, not being the same in all directions of propagation.
NRZ	Non-return to zero, a data encoding scheme.
ns	Nanosecond; one billionth of a second.
Nyquist frequency	In sampling theory, this is the highest frequency that can be faithfully reproduced, equal to two times the system resolution. Also called the Nyquist criterion.
NZDSF	Non-zero dispersion-shifted fiber; a fiber made to operate in the 1550-nm window so that effects such as Four-wave mixing are minimized. ITU-T: G.655 has a little bit of CD in 1550 nm; some of them have also a reduced CD slope in the 1550 nm.
obscuration	An opaque light blockage in the optics of a system.
OC	Optical channel.
OEO	Optical to electronic to optical. Generally used to refer to the process of converting an optical signal to an electronic signal for some function or process, then converting it back to an optical signal.
OFA	Optical fiber amplifier, generally an erbium-doped amplifier (EDFA).

OFC	Optical fiber conference, a major conference on optical telecom held each spring.
OOO	Optical to optical; generally used to refer to an all-optical component that doesn't require conversion of the optical signal to an electronic signal.
OPD	Optical path difference. Used in describing the performance limitations of optical systems in which the phase front of the light propagating in the system is aberrated by the optics or the turbulent environment outside of the optics.
optical depth	The integration of the absorption coefficient along the path of light in an absorbing medium.
optical element	A lens, mirror, flat, or other piece of optics used in an optical assembly.
opto-mechanics	The field of engineering that addresses the integration of optics and mechanical structures and mechanisms.
ORL	Optical return loss.
OSA	Optical spectrum analyzer or Optical Society of America.
OTDR	Optical time domain reflectometry, a technique used to locate breaks in a fiber by the back-reflection caused by the break.
OXC	Optical cross connection.
PC	Photoconductive.
PCB	Printed circuit board.
PCR	Peak cell rate.
PDG	Polarization-dependent gain.
PDL	Polarization-dependent loss.
PDR	Preliminary design review. A meeting at which early plans for system development are presented and reviewed.
Peltier	The discoverer of the basic physics that led to the development of the thermoelectric cooler.
Pfa	Probability of false alarm. A term often found in discussions of detection theory and the design and performance assessment of such systems.
PHB	Polarization hole burning.

photoconductive	A type of detector that changes conductance (resistance) on exposure to light. Contrasts with the photovoltaic detector types.
photodetectors	A broad description of components that are able to sense the presence of light in a quantitative way. Generally, this category includes photomultipliers, photodiodes, CCDs, etc.
photo-emission	A process of the emission of an electron from a surface when a photon encounters it. This is the operational basis of the photomultiplier tube.
photolithographic	A process using masks and etching to create very minute (less than a micron is common) features. It is widely used by the electronics industry to create integrated circuits.
photon	A massless bundle of energy traveling at the speed of light and associated with light and light phenomena.
photonics	The general field of creation, detection, and manipulation of photons of light. Includes virtually all of the modern electro-optic sciences.
photopic	Characterizing the response of the cones in the human retina. See also *scotopic.*
photovoltaic	A type of detector that produces an output voltage difference between two electrodes in response to impinging light.
PIC	Photonic integrated circuit.
pico	Referring to 10^{-12}, as in "a picowatt equals one-millionth of one-millionth of a watt."
PIN	Positive-intrinsic-negative (a type of photodiode).
pixel	Traditionally, a picture element, and the term should be strictly applied only to displays. However, it is sometimes used to refer to an individual detector on a focal plane array.
Planck function	Refers to Max Planck's blackbody radiation law.
PLC	Planar lightwave component or circuit.
PMD	Polarization mode dispersion.
POF	Plastic optical fibers.
pointance	See *luminous pointance.*
PON	Passive optical network.

POP	Point of presence.
producibility	The measure of the ability and effort required to produce a particular item (system or component) at a given production rate.
projected area	The effective two-dimensional silhouette of a three-dimensional object. A sphere has a projected area of a circle of the same radius.
PSD	Power spectral density.
PV	When used with detectors, *photovoltaic*. (See definition above.) When used with optics, *peak to valley*.
QAM	Quadrature amplitude modulation.
quantum efficiency	When used in reference to detectors, this is a figure of merit describing the conversion efficiency from a photon to a usable free carrier. A 100% quantum efficient implies that all the photons incident generate carriers that can be read out.
QWIP	Quantum well infrared photodetector (or photoconductor).
RC	Resistance-capacitance, generally the product of a device's resistance times its capacitance.
radiant areance	watts/meter^2.
radiant pointance	watts/steradian.
radiant sterance	watts/meter^2/steradian.
radiometry	The study of the emission, transmission, and absorptance of radiant energy. Some of the entries in this glossary define basic parameters of radiometry. See, for example, *pointance, sterance,* and *areance.* See also any number of on-line resources such as the NIST web site (http://www.nist.gov) or http://www.optics.arizona.edu/Palmer/rpfaq/rpfaq.htm for more detailed descriptions.
rangefinder	Active or passive optical systems capable of measuring the distance between points. A laser radar (LADAR) performs this function by illuminating the target and measuring the time delay until the reflected light appears at a sensor co-located with the transmitter. Many other systems can be employed to make range-finding measurements.

Rayleigh scattering	A scattering process named after the theory developed by Lord Rayleigh (John William Strutt) to describe the scattering properties of the atmosphere. Generally, Raleigh scattering effect decreases as wavelength increases (which is why the sky is blue and not red).
repeater	A telecommunications subsystem that receives a telecommunications signal, amplifies and/or conditions it, and rebroadcasts it.
RF	Radio frequency.
RIN	Relative intensity noise.
r_o	A common abbreviation for Fried's parameter, important to free-space telecom.
router	A device that determines the best available route to send a packet and instructs the switches accordingly.
RZ	Return to zero.
SAP	Service access point.
SBS	Stimulated Brillouin scattering.
SDH	Synchronous digital hierarchy.
semiconductor	A material with a concentration of free carriers, which allows it to be a conductor or resistor, and allows this property to be controlled by an applied voltage.
Si	Silicon. The most commonly used material in the development of semiconductors. Also used as a material for visible and near-infrared detectors and, when doped, short-, mid-, and long-wavelength infrared focal planes. Truly nature's gift to us all.
SIA	Semiconductor Industry Association.
SMF	Single-mode fiber.
SNR	Signal-to-noise ratio.
SOA	Semiconductor optical amplifier or state of the art.
SOI	Silicon on insulator, an integrated circuit architecture that separates the active electronic integrated circuit from the bulk silicon, yielding functions requiring fewer electrons, resulting in lower power, faster speeds, and less heat generation.
soliton	A shaped "phase-conjugation" waveform meant to take advantage of the fiber's nonlinear properties to project pulses over great distances with little to no

loss; thus, a pulse bundle that exploits dispersion and self-phase modulation. Wavelengths at the leading edge of the pulse effectively travel more slowly, and the wavelengths at the trailing edge effectively travel faster.

SONET   Synchronous optical network.

SPIE   Society of Photo-Optical Instrumentation Engineers, an international professional society that deals with all of the subjects pertaining to this book, and many others (www.spie.org).

SRR   System requirements review. An early meeting in the development of a system, during which the requirements for the system, derived from performance requirements, can be reviewed and improved.

step-index fibers   A fiber in which there is a discrete change in the index of refraction between the core and cladding.

steradian   The unit of measure of solid angle. Computed as the ratio of the area of the cap of a sphere to the square of the radius to the point at which the solid angle is measured.

sterance   See *radiant sterance* and *luminous sterance*.

STM   Synchronous transfer mode; generally a mode in which there is a constant temporal interval between adjacent bits. It therefore does not include commands to signify the beginning and end of a transmission.

Strehl   An optics researcher who lends his name to the oft-quoted measure of the performance of an optical system in which the intensity that is projected or collected is compared with that of a perfect system.

STS   Synchronous transport signal.

subtense   The angular extent of a system, such as the field of view of a sensor.

synchronous digital hierarchy   A high-speed SONET protocol transmitting up to 40 Gb/s.

TBS   Terabits per second; sometimes, terabytes per second.

TCM   Trellis code modulation.

TDM   Time division multiplexing.

TEC   Thermoelectric cooler.

TICL	Temperature-induced cable loss.
TIS	Total integrated scatter.
torr	A measure of pressure equal to 133.32 pascals.
transmission	The measure of the amount of radiation that passes through a substance.
transmissive	Having the property of being partially transparent. Must be specified as to wavelength.
transmittance	The ratio of the amount of radiation of a particular wavelength that passes through a path in a material to the amount incident on the material.
TTL	Transistor-to-transistor logic (usually operating at 3 to 5 V).
unitless	Having properties such that the parameter is not associated with any particular measure of distance, volume, time, etc.
UPSR	Unidirectional path switched ring.
UV	Ultraviolet. This is part of the electromagnetic spectrum with wavelengths longer than X-rays (≈ 0.1 μm) but shorter than visible light (≈ 0.4 μm).
VAD	Vapor axial deposition or vapor phase axial deposition.
VBR	Variable bit rate.
VC	Venture capital.
VCC	Virtual channel connection.
VCSEL	Vertical-cavity surface emitting laser, generally a quantum well laser operating in the near IR and emitting vertically from the integrated circuit.
VDSL	Very high-speed digital subscriber line.
visible spectrum	The part of the electromagnetic spectrum that is approximately visible to the human eye. Generally, this is from about 0.35 to 0.76 μm.
VME	Versa module Europe. A standardized circuit board size and I/O architecture, popular with military systems.
VP	Virtual path.
VPC	Virtual path connection.
VT	Virtual tributary.
wafer	A slice of a boule, usually used to delineate a material disk that is suitable for semiconductor processing.

WAN	Wide area network.
wave	A periodic undulation in a field.
wave number	A measure of the frequency of electromagnetic radiation. Computed by the formula $1/\lambda$.
wavefront	Characterization of a beam of light, usually for the purpose of characterizing the degradations induced by passage through an imperfect optical system or transmission medium.
waveguide	A pipe or other conveyance able to efficiently aid the transmission of light or infrared radiation from one point to another. Optical fibers meet this definition.
wavelength	The distance from one peak in a field of waves to the next.
WDM	Wavelength division multiplexing, the practice of placing multiple simultaneous telecom wavelengths in a single fiber; generally less than 10 wavelengths (if more, it is termed "DWDM").
WFE	Wavefront error, such as describes the degraded properties of light propagating through imperfect optics.
WGR	Waveguide grating router,
white noise	Noise that is not frequency dependent (has a flat power spectral density).
WIC	Wavelength-independent coupler.
Wide area network	A network linking telecommunications over a larger area than a single work site or metropolitan area, often based on X.25 packet switching.
XPM	Cross phase modulation; a destructive interaction between different phases of the light of a pulse. This is a nonlinear effect of a fiber.
ZBLAN	A fluoride glass type of fiber that has low loss from the near infrared through mid-wave infrared.

Index

About the Authors

John Lester Miller earned a B.S. in Physics at the University of Southern California in 1981; participated in physics, math, and engineering graduate studies at Cal State Long Beach and the University of Hawai'i; then earned an M.B.A. in technology management from Regis University in 1989. He chairs the SPIE session on advanced infrared technology and frequently referees papers for electro-optical journals. He is now a senior program director and general manager of the Portland office of the Research Triangle Institute. Previously, he has held positions as Chief Scientist, Director of Advanced Technologies, Program Manager, Functional Manager, Lead Engineer, and Electro-Optical Engineer with FLIR Systems (Portland), Lockheed Martin (Denver, Utica, and Orlando), University of Hawaii's NASA IRTF (Hilo), Rockwell International (Seal Beach), and Mt. Wilson and Palomar Observatories (Pasadena) and Griffith Observatory (Los Angeles). He has published more than 30 papers on optical sciences and is author of *Principles of Infrared Technology* and the co-author of the previous edition of this book. He has several patents pending in optical technologies.

John's experience includes leading efforts on telecom optical performance monitors, laser communication systems, surveillance systems, cameras, laser imaging systems, fiber optics, environmental and weather monitoring sensors, and image processing. John was the technical lead for FLIR Systems' telecom endeavors. John and His wife, Corinne Foster, split their time between Lake Oswego and Bend Oregon.

Ed Friedman earned a B.S. in Physics at the University of Maryland in 1966 and a Ph.D. in cryogenic physics from Wayne State University in 1972. He started his career in the field of ocean optics and subsequently developed system concepts for remote sensing of the atmosphere and climate. After completing studies related to the design of spacecraft and instruments for the measurement of the radiation balance of the Earth, he was appointed a visiting scientist in the climate program at the National Center for Atmospheric Research in Boulder, Colorado. Subsequent employers included The Mitre Corporation, Martin Marietta, Ball Aerospace and Technologies Corporation, and the The Boeing Company, where he serves as a Technical Fellow in the Lasers and Electro-Optics Division.

In the last eight years, he has concentrated on the development of mission concepts and technologies for astrophysics and space science. While at Ball Aerospace and Technologies Corporation, he was Chief Technologist of the Civil Space business unit. Recent areas of interest include the use of space-based interferometers to create detailed maps of stellar positions and the use of coronagraphic methods for detection of planets in distant solar systems. He and his Ph.D. student were recently awarded a patent for a novel method of alignment and phasing of large, deployed Earth-viewing optics. He has been a patent reviewer for the journal *Applied Optics* and an editor for the journal *Optical Engineering*. He recently retired after ten seasons as a member of the National Ski Patrol. He and his wife Judith live in the mountains west of Boulder, Colorado.